Springer Uncertainty Research

Springer Uncertainty Research

Springer Uncertainty Research is a book series that seeks to publish high quality monographs, texts, and edited volumes on a wide range of topics in both fundamental and applied research of uncertainty. New publications are always solicited. This book series provides rapid publication with a world-wide distribution.

Editor-in-Chief
Baoding Liu
Department of Mathematical Sciences
Tsinghua University
Beijing 100084, China
http://orsc.edu.cn/liu
Email: liu@tsinghua.edu.cn

Executive Editor-in-Chief
Kai Yao
School of Economics and Management
University of Chinese Academy of Sciences
Beijing 100190, China
http://orsc.edu.cn/yao
Email: yaokai@ucas.ac.cn

More information about this series at http://www.springer.com/series/13425

Kai Yao

Uncertain Differential Equations

 Springer

Kai Yao
School of Economics and Management
University of Chinese Academy of Sciences
Beijing
China

ISSN 2199-3807   ISSN 2199-3815   (electronic)
Springer Uncertainty Research
ISBN 978-3-662-57075-3   ISBN 978-3-662-52729-0   (eBook)
DOI 10.1007/978-3-662-52729-0

Printed on acid-free paper

This Springer imprint is published by Springer Nature
The registered company is Springer-Verlag GmbH Berlin Heidelberg

To my parents
Yuesheng Yao
Xiuying Zhang

Preface

Uncertainty theory is a branch of mathematics for modeling belief degrees. Within the framework of uncertainty theory, uncertain variable is used to represent quantities with uncertainty, and uncertain process is used to model the evolution of uncertain quantities. Uncertain differential equation is a type of differential equations involving uncertain processes. Since it was proposed in 2008, uncertain differential equation has been subsequently studied by many researchers. So far, it has become the main tool to deal with dynamic uncertain systems.

Uncertain Variable

Uncertain measure is used to quantify the belief degree that an uncertain event is supposed to occur, and uncertain variable is used to represent quantities with human uncertainty. Chapter 2 is devoted to uncertain measure, uncertain variable, uncertainty distribution, inverse uncertainty distribution, operational law, expected value, and variance.

Uncertain Process

Uncertain process is essentially a sequence of uncertain variables indexed by the time. Chapter 3 introduces some basic concepts about an uncertain process, including uncertainty distribution, extreme value, and time integral.

Contour Process

Contour process is a type of uncertain processes with some special structures so that its main properties are determined by a spectrum of its sample paths. Solutions of uncertain differential equations are the most frequently used contour processes. Chapter 4 is devoted to such processes and proves the set of contour processes is closed under the extreme value operator, time integral operator, and monotone function operator.

Uncertain Calculus

Uncertain calculus deals with the differentiation and integration of uncertain processes. Chapter 5 introduces the Liu process, the Liu integral, the fundamental theorem, and integration by parts.

Uncertain Differential Equation

Uncertain differential equation is a type of differential equations involving uncertain processes. Chapter 6 is devoted to the uncertain differential equations driven by the Liu processes. It discusses some analytic methods and numerical methods for solving uncertain differential equations. In addition, the existence and uniqueness theorem, and stability theorems on the solution of an uncertain differential equation are also covered. For application, it introduces two stock models and derives their option pricing formulas as well.

Uncertain Calculus with Renewal Process

Renewal process is a type of discontinuous uncertain processes, which is used to record the number of renewals of an uncertain system. Chapter 7 is devoted to uncertain calculus with respect to renewal process. It introduces the renewal process, the Yao integral and the Yao process, including the fundamental theorem and integration by parts.

Uncertain Differential Equation with Jumps

Uncertain differential equation with jumps is essentially a type of differential equations driven by both the Liu processes and the renewal processes. Chapter 8 is devoted to uncertain differential equation with jumps, including the existence and uniqueness, and stability of its solution. It also introduces a stock model with jumps and derives its option pricing formulas for application purpose.

Multi-Dimensional Uncertain Differential Equation

Multi-dimensional uncertain differential equation is a system of uncertain differential equations. Chapter 9 introduces multi-dimensional Liu process, multi-dimensional uncertain calculus, and multi-dimensional uncertain differential equation.

High-Order Uncertain Differential Equation

High-order uncertain differential equation is a type of differential equations involving the high-order derivatives of uncertain processes. Chapter 10 is devoted to high-order uncertain differential equations driven by the Liu processes. It gives a numerical method for solving high-order uncertain differential equations. In addition, the existence and uniqueness theorem on the solution of a high-order uncertain differential equation is also covered.

Uncertainty Theory Online

If you would like to read more papers related to uncertain differential equations, please visit the Web site at http://orsc.edu.cn/online.

Purpose

The purpose of this book was to provide a tool for handling dynamic systems with human uncertainty. The book is suitable for researchers, engineers, and students in the field of mathematics, information science, operations research, industrial engineering, economics, finance, and management science.

Acknowledgment

This work was supported in part by National Natural Science Foundation of China (Grant No.61403360). I would like to express my sincere gratitude to Prof. Baoding Liu of Tsinghua University for his rigorous supervision. My sincere thanks also go to Prof. Jinwu Gao of Renmin University of China, Prof. Xiaowei Chen of Nankai Univeristy, Prof. Ruiqing Zhao of Tianjin University, Prof. Yuanguo Zhu of Nanjing University of Science and Technology, and Prof. Jin Peng of Huanggang Normal University. I am also deeply grateful to my wife, Meixia Wang, for her love and support.

Beijing   Kai Yao
February 2016

Contents

Frequently Used Symbols

$\mathcal{M}$	Uncertain measure
$(\Gamma, \mathcal{L}, \mathcal{M})$	Uncertainty space
ξ, η, τ	Uncertain variables
$\boldsymbol{\xi}$	Uncertain vector
Φ, Ψ, Υ	Uncertainty distributions
$\Phi^{-1}, \Psi^{-1}, \Upsilon^{-1}$	Inverse uncertainty distributions
$\mathcal{L}(a, b)$	Linear uncertain variable
$\mathcal{N}(e, \sigma)$	Normal uncertain variable
$\mathcal{LOGN}(e, \sigma)$	Lognormal uncertain variable
E	Expected value
V	Variance
X_t, Y_t, Z_t	Uncertain processes
C_t	Canonical Liu process
N_t	Uncertain renewal process
$\boldsymbol{X}_t, \boldsymbol{Y}_t, \boldsymbol{Z}_t$	Multi-dimensional uncertain processes
$\boldsymbol{C}_t$	Multi-dimensional canonical Liu process
$\bigvee$	Maximum operator
$\bigwedge$	Minimum operator

Chapter 1
Introduction

Uncertain differential equation is a type of differential equations involving uncertain processes. So far, uncertain differential equation driven by the Liu process, uncertain differential equation with jumps, multi-dimensional uncertain differential equation, and high-order uncertain differential equation have been proposed, which are all covered in this book.

1.1 Uncertain Differential Equation

The concept of uncertain differential equation was first proposed in 2008, when Liu [37] presented a type of differential equations driven by the Liu processes, which are usually called uncertain differential equations for simplicity. In 2010, Chen and Liu [3] found the analytic solution of a linear uncertain differential equation. After that, Liu [52] and Yao [77] proposed analytic methods for solving two special types of uncertain differential equations, which were generalized by Liu [50] and Wang [87] later. Note that the analytic methods do not always work, for the analytic solution may have a very complex form or not exist at all. Therefore, numerical methods play the core role in solving uncertain differential equations. In 2013, Yao and Chen [75] found that the solution of an uncertain differential equation can be represented by the solutions of a spectrum of ordinary differential equations. Later, Yao [76] found the extreme value and the time integral of the solution of an uncertain differential equation could also be represented by the solutions of these ordinary differential equations. Thus instead of solving an uncertain differential equation, we only need to solve these ordinary uncertain differential equations. Yao and Chen [75] suggested to solve the ordinary differential equations via Euler scheme, while Yang and Shen [69] recommended the Runge–Kutta scheme.

Other than the various methods for solving uncertain differential equations, properties of the solution of an uncertain differential equation have also been widely investigated. Chen and Liu [3] demonstrated that an uncertain differential equation has a unique solution if its coefficients satisfy the linear growth condition and the

© Springer-Verlag Berlin Heidelberg 2016
K. Yao, *Uncertain Differential Equations*,
Springer Uncertainty Research, DOI 10.1007/978-3-662-52729-0_1

Lipschitz condition. Later, Gao [19] showed the Lipschitz condition can be weakened to the local Lipschitz condition. Stability of a differential equation means that a perturbation on the initial value will not result in an influential shift. So far, there are mainly three types of stability for an uncertain differential equation, namely stability in measure, stability in mean and almost sure stability. The concept of stability in measure was proposed by Liu [38], and a sufficient condition was given by Yao et al. [74]. Stability in mean and almost surely stability were studied by Yao et al. [82] and Liu et al. [49], respectively. As a generalization of stability in mean, Sheng and Wang [60] proposed a concept of stability in p-th moment. In addition, Gao and Yao [20] studied the continuous dependence of the solution on the initial value.

Uncertain differential equation finds a variety of applications in finance. Liu [38] supposed the stock price follows an uncertain differential equation, proposed the first uncertain stock model (which is usually called Liu's stock model), and derived its European option pricing formulas. Then Chen [4] and Sun and Chen [63] derived its American and Asian option pricing formulas, respectively. In addition, Yao [80] found a sufficient and necessary no-arbitrage condition for Liu's stock model. Inspired by Liu's stock model, Peng and Yao [55] and Chen et al. [9] presented mean-reverting stock model and periodic-dividend stock model, respectively. Besides, Yao [85] provided an uncertain stock model with floating interest rate. Uncertain interest rate model was first proposed by Chen and Gao [10], and a numerical method to calculate the price of its zero-coupon bond was designed by Jiao and Yao [32]. Uncertain currency model was proposed by Liu et al. [53].

Uncertain differential equation has also been introduced to optimal control. Based on the principle of optimality, Zhu [93] derived an optimality equation of uncertain control model, which was further developed by Zhu and his coauthors (Ge and Zhu [22, 23], Sheng and Zhu [59]).

1.2 Uncertain Differential Equation with Jumps

The concept of uncertain differential equation with jumps was first proposed by Yao [72] in 2012. It is essentially a type of differential equations driven by the canonical Liu processes and the uncertain renewal processes. The solutions of two special types of uncertain differential equations with jumps were given by Yao [83]. So far, a sufficient condition for an uncertain differential equation with jumps having a unique solution has been given by Yao [83], and concepts of stability in measure and almost sure stability for an uncertain differential equation with jumps have been studied by Yao [83] and Ji and Ke [29], respectively. In addition, Yu [88] assumed the stock price follows an uncertain differential equation with jumps and proposed an uncertain stock model with jumps, which was generalized by Ji and Zhou [30] later.

1.3   Multi-Dimensional Uncertain Differential Equation

The concept of multi-dimensional uncertain differential equation was first proposed in 2014, when Yao [79] presented a multi-dimensional differential equation driven by a multi-dimensional Liu process. Essentially, a multi-dimensional uncertain differential equation is a system of uncertain differential equations. The solutions of two special types of linear multi-dimensional uncertain differential equations were given by Ji and Zhou [31]. Besides, a sufficient condition for a multi-dimensional uncertain differential equation having a unique solution was given by Ji and Zhou [31], and a concept of stability in measure for a multi-dimensional uncertain differential equation by Su et al. [62].

1.4   High-Order Uncertain Differential Equation

High-order uncertain differential equation is a type of differential equations involving the high-order derivatives of uncertain processes. Essentially, it can be transformed into a multi-dimensional uncertain differential equation. This book shows that the solution of a high-order uncertain differential equation can be represented by the solutions of a spectrum of high-order ordinary differential equations, based on which a numerical method is designed to solve a high-order uncertain differential equation.

Chapter 2
Uncertain Variable

Uncertainty theory was founded by Liu [36] in 2007 and perfected by Liu [38] in 2009 to deal with human's belief degree based on four axioms, and uncertain variable is the main tool to model a quantity with human uncertainty in uncertainty theory. The emphases of this chapter are on uncertain measure, uncertain variable, uncertainty distribution, inverse uncertainty distribution, operational law, expected value, and variance.

2.1 Uncertain Measure

Let $\mathcal{L}$ be a σ-algebra on a nonempty set Γ. Then each element $\Lambda \in \mathcal{L}$ is called an event. Uncertain measure $\mathcal{M}$ is a function from $\mathcal{L}$ to $[0, 1]$, that is, it assigns to each event Λ a number $\mathcal{M}\{\Lambda\}$ which indicates the belief degree that the event Λ will occur. According to the properties of belief degree, Liu [36] proposed the following three axioms that an uncertain measure is supposed to satisfy:

Axiom 1 (*Normality Axiom*) $\mathcal{M}\{\Gamma\} = 1$ for the universal set Γ.

Axiom 2 (*Duality Axiom*) $\mathcal{M}\{\Lambda\} + \mathcal{M}\{\Lambda^c\} = 1$ for any event Λ.

Axiom 3 (*Subadditivity Axiom*) For every countable sequence of events Λ_1, $\Lambda_2, \ldots$, we have

$$\mathcal{M}\left\{\bigcup_{i=1}^{\infty} \Lambda_i\right\} \leq \sum_{i=1}^{\infty} \mathcal{M}\{\Lambda_i\}.$$

Definition 2.1 (*Liu [36]*) A set function $\mathcal{M}$ on a σ-algebra $\mathcal{L}$ of a nonempty set Γ is called an uncertain measure if it satisfies the normality, duality, and subadditivity axioms. In this case, the triple $(\Gamma, \mathcal{L}, \mathcal{M})$ is called an uncertainty space.

© Springer-Verlag Berlin Heidelberg 2016
K. Yao, *Uncertain Differential Equations*,
Springer Uncertainty Research, DOI 10.1007/978-3-662-52729-0_2

Example 2.1 Let $\mathcal{L}$ be the power set of $\Gamma = \{\gamma_1, \gamma_2, \gamma_3\}$. Define

$$\mathcal{M}\{\gamma_1\} = 0.5, \quad \mathcal{M}\{\gamma_2\} = 0.4, \quad \mathcal{M}\{\gamma_3\} = 0.3,$$

$$\mathcal{M}\{\gamma_1, \gamma_2\} = 0.7, \quad \mathcal{M}\{\gamma_1, \gamma_3\} = 0.6, \quad \mathcal{M}\{\gamma_2, \gamma_3\} = 0.5,$$

$$\mathcal{M}\{\emptyset\} = 0, \quad \mathcal{M}\{\Gamma\} = 1.$$

Then $\mathcal{M}$ is an uncertain measure, and $(\Gamma, \mathcal{L}, \mathcal{M})$ is an uncertainty space.

Example 2.2 Let $\lambda(x)$ be a nonnegative function on $\Re$ satisfying

$$\sup_{x \neq y}(\lambda(x) + \lambda(y)) = 1,$$

and $\mathcal{B}$ be the Borel algebra of $\Re$. For each Borel set $B \in \mathcal{B}$, define

$$\mathcal{M}\{B\} = \begin{cases} \sup_{x \in B} \lambda(x), & \text{if } \sup_{x \in B} \lambda(x) < 0.5 \\ 1 - \sup_{x \in B^c} \lambda(x), & \text{if } \sup_{x \in B} \lambda(x) \geq 0.5. \end{cases}$$

Then $\mathcal{M}$ is an uncertain measure, and $(\Re, \mathcal{B}, \mathcal{M})$ is an uncertainty space.

Example 2.3 Let $\rho(x)$ be a nonnegative and integrable function on $\Re$ satisfying

$$\int_{\Re} \rho(x)\mathrm{d}x \geq 1,$$

and $\mathcal{B}$ be the Borel algebra of $\Re$. For each Borel set $B \in \mathcal{B}$, define

$$\mathcal{M}\{B\} = \begin{cases} \int_B \rho(x)\mathrm{d}x, & \text{if } \int_B \rho(x)\mathrm{d}x < 0.5 \\ 1 - \int_{B^c} \rho(x)\mathrm{d}x, & \text{if } \int_{B^c} \rho(x)\mathrm{d}x < 0.5 \\ 0.5, & \text{otherwise.} \end{cases}$$

Then $\mathcal{M}$ is an uncertain measure, and $(\Re, \mathcal{B}, \mathcal{M})$ is an uncertainty space.

Theorem 2.1 (Liu [40], Monotonicity Theorem) *For any events $\Lambda_1 \subset \Lambda_2$, we have*

$$\mathcal{M}\{\Lambda_1\} \leq \mathcal{M}\{\Lambda_2\}. \tag{2.1}$$

Proof Since $\Lambda_1 \subset \Lambda_2$, we have $\Gamma = \Lambda_1^c \cup \Lambda_2$. By using the subadditivity and duality of uncertain measure, we obtain

$$1 = \mathcal{M}\{\Gamma\} \leq \mathcal{M}\{\Lambda_1^c\} + \mathcal{M}\{\Lambda_2\} = 1 - \mathcal{M}\{\Lambda_1\} + \mathcal{M}\{\Lambda_2\}.$$

Hence $\mathcal{M}\{\Lambda_1\} \leq \mathcal{M}\{\Lambda_2\}$. The theorem is proved.

Theorem 2.2 (Liu [36]) *For any event Λ, we have*

$$0 \leq \mathcal{M}\{\Lambda\} \leq 1. \tag{2.2}$$

Proof Since $\emptyset \subset \Lambda \subset \Gamma$, $\mathcal{M}\{\Gamma\} = 1$, and $\mathcal{M}\{\emptyset\} = 1 - \mathcal{M}\{\Gamma\} = 0$, we have $0 \leq \mathcal{M}\{\Lambda\} \leq 1$ according to the monotonicity of uncertain measure. The theorem is proved.

Product Uncertain Measure

Let $(\Gamma_k, \mathcal{L}_k, \mathcal{M}_k)$, $k = 1, 2, \ldots$ be a sequence of uncertainty spaces. Write $\Gamma = \Gamma_1 \times \Gamma_2 \times \cdots$ as the universal set, and $\mathcal{L} = \mathcal{L}_1 \times \mathcal{L}_2 \times \cdots$ as the product σ-algebra. In 2009, Liu [38] defined an uncertain measure $\mathcal{M}$ on $\mathcal{L}$, producing the fourth axiom of uncertain measure.

Axiom 4 (*Product Axiom*) Let $(\Gamma_k, \mathcal{L}_k, \mathcal{M}_k)$ be uncertainty spaces for $k = 1, 2, \ldots$. The product uncertain measure $\mathcal{M}$ is an uncertain measure satisfying

$$\mathcal{M}\left\{\prod_{k=1}^{\infty} \Lambda_k\right\} = \bigwedge_{k=1}^{\infty} \mathcal{M}_k\{\Lambda_k\}$$

where Λ_k are arbitrarily chosen events from $\mathcal{L}_k$ for $k = 1, 2, \ldots$, respectively.

For an arbitrary event $\Lambda \in \mathcal{L}$, its uncertain measure could be obtained via

$$\mathcal{M}\{\Lambda\} = \begin{cases} \sup\limits_{\Lambda_1 \times \Lambda_2 \times \cdots \subset \Lambda} \min\limits_{1 \leq k < \infty} \mathcal{M}_k\{\Lambda_k\}, \\ \qquad \text{if} \quad \sup\limits_{\Lambda_1 \times \Lambda_2 \times \cdots \subset \Lambda} \min\limits_{1 \leq k < \infty} \mathcal{M}_k\{\Lambda_k\} > 0.5 \\ 1 - \sup\limits_{\Lambda_1 \times \Lambda_2 \times \cdots \subset \Lambda^c} \min\limits_{1 \leq k < \infty} \mathcal{M}_k\{\Lambda_k\}, \\ \qquad \text{if} \quad \sup\limits_{\Lambda_1 \times \Lambda_2 \times \cdots \subset \Lambda^c} \min\limits_{1 \leq k < \infty} \mathcal{M}_k\{\Lambda_k\} > 0.5 \\ 0.5, \qquad \text{otherwise.} \end{cases}$$

Peng and Iwamura [58] showed that the triple $(\Gamma, \mathcal{L}, \mathcal{M})$ derived as above from the uncertainty spaces $(\Gamma_k, \mathcal{L}_k, \mathcal{M}_k)$, $k = 1, 2, \ldots$ is an uncertainty space. Interested readers may refer to Peng and Iwamura [58] or Liu [40] for details.

2.2 Uncertain Variable

Definition 2.2 *(Liu [36])* An uncertain variable ξ is a measurable function from an uncertainty space $(\Gamma, \mathcal{L}, \mathcal{M})$ to the set of real numbers, i.e., for any Borel set B of real numbers, the set

$$\{\xi \in B\} = \{\gamma \in \Gamma \mid \xi(\gamma) \in B\} \tag{2.3}$$

is an event.

Example 2.4 Consider an uncertainty space $(\Gamma, \mathcal{L}, \mathcal{M})$ with $\Gamma = \{\gamma_1, \gamma_2, \gamma_3\}$ and $\mathcal{M}\{\gamma_1\} = 0.5, \mathcal{M}\{\gamma_2\} = 0.4, \mathcal{M}\{\gamma_3\} = 0.3$. The function ξ defined by

$$\xi(\gamma) = \begin{cases} -1, & \text{if } \gamma = \gamma_1 \\ 0, & \text{if } \gamma = \gamma_2 \\ 1, & \text{if } \gamma = \gamma_3 \end{cases}$$

is an uncertain variable on $(\Gamma, \mathcal{L}, \mathcal{M})$.

Theorem 2.3 (Liu [36]) *Suppose f is a measurable function, and $\xi_1, \xi_2, \ldots, \xi_n$ are uncertain variables on $(\Gamma, \mathcal{L}, \mathcal{M})$. Then the function*

$$\xi = f(\xi_1, \xi_2, \ldots, \xi_n) \tag{2.4}$$

defined by

$$\xi(\gamma) = f(\xi_1(\gamma), \xi_2(\gamma), \ldots, \xi_n(\gamma)), \quad \forall \gamma \in \Gamma \tag{2.5}$$

is an uncertain variable.

Proof Since $\xi_1, \xi_2, \ldots, \xi_n$ are measurable functions from $(\Gamma, \mathcal{L}, \mathcal{M})$ to the set of real numbers, and f is a measurable function on the set of real numbers, the composite function $f(\xi_1, \xi_2, \ldots, \xi_n)$ is also a measurable function from $(\Gamma, \mathcal{L}, \mathcal{M})$ to the set of real numbers. Hence ξ is an uncertain variable. The theorem is proved.

Example 2.5 Let ξ_1 and ξ_2 be two uncertain variables. Then the sum $\eta = \xi_1 + \xi_2$ defined by

$$\eta(\gamma) = \xi_1(\gamma) + \xi_2(\gamma), \quad \forall \gamma \in \Gamma$$

is an uncertain variable, and the product $\tau = \xi_1 \cdot \xi_2$ defined by

$$\tau(\gamma) = \xi_1(\gamma) \cdot \xi_2(\gamma), \quad \forall \gamma \in \Gamma$$

is also an uncertain variable.

Independence

Definition 2.3 (*Liu [38]*) The uncertain variables $\xi_1, \xi_2, \ldots, \xi_n$ are said to be independent if

$$\mathcal{M}\left\{\bigcap_{i=1}^{n}(\xi_i \in B_i)\right\} = \bigwedge_{i=1}^{n} \mathcal{M}\{\xi_i \in B_i\} \tag{2.6}$$

for any Borel sets $B_1, B_2, \ldots, B_n$ of real numbers.

Theorem 2.4 (Liu [38]) *The uncertain variables $\xi_1, \xi_2, \ldots, \xi_n$ are independent if and only if*

$$\mathcal{M}\left\{\bigcup_{i=1}^{n}(\xi_i \in B_i)\right\} = \bigvee_{i=1}^{n} \mathcal{M}\{\xi_i \in B_i\} \tag{2.7}$$

for any Borel sets $B_1, B_2, \ldots, B_n$ of real numbers.

Proof On the one hand, suppose that $\xi_1, \xi_2, \ldots, \xi_n$ are independent uncertain variables. Then we have

$$\mathcal{M}\left\{\bigcup_{i=1}^{n}(\xi_i \in B_i)\right\} = 1 - \mathcal{M}\left\{\bigcap_{i=1}^{n}(\xi_i \in B_i^c)\right\}$$

$$= 1 - \bigwedge_{i=1}^{n} \mathcal{M}\{\xi_i \in B_i^c\} = \bigvee_{i=1}^{n} \mathcal{M}\{\xi_i \in B_i\}.$$

Thus Eq. (2.7) holds. On the other hand, suppose that Eq. (2.7) holds. Then we have

$$\mathcal{M}\left\{\bigcap_{i=1}^{n}(\xi_i \in B_i)\right\} = 1 - \mathcal{M}\left\{\bigcup_{i=1}^{n}(\xi_i \in B_i^c)\right\}$$

$$= 1 - \bigvee_{i=1}^{n} \mathcal{M}\{\xi_i \in B_i^c\} = \bigwedge_{i=1}^{n} \mathcal{M}\{\xi_i \in B_i\}.$$

Hence the uncertain variables $\xi_1, \xi_2, \ldots, \xi_n$ are independent. The theorem is proved.

Uncertain Vector

Definition 2.4 (*Liu [36]*) Let $\xi_1, \xi_2, \ldots, \xi_m$ be some uncertain variables. Then the vector

$$\xi = (\xi_1, \xi_2, \ldots, \xi_m) \tag{2.8}$$

is called an m-dimensional uncertain vector.

The concept of independence for uncertain vectors is a generalization of that for uncertain variables.

Definition 2.5 (*Liu [45]*) The m-dimensional uncertain vectors $\xi_1, \xi_2, \ldots, \xi_n$ are said to be independent if

$$\mathcal{M}\left\{\bigcap_{i=1}^{n} (\xi_i \in B_i)\right\} = \bigwedge_{i=1}^{n} \mathcal{M}\{\xi_i \in B_i\} \tag{2.9}$$

for any Borel sets $B_1, B_2, \ldots, B_n$ of m-dimensional real vectors.

2.3 Uncertainty Distribution

Definition 2.6 (*Liu [36]*) Let ξ be an uncertain variable. Then its uncertainty distribution Φ is defined by

$$\Phi(x) = \mathcal{M}\{\xi \leq x\}, \quad \forall x \in \Re. \tag{2.10}$$

Example 2.6 Consider an uncertainty space $(\Gamma, \mathcal{L}, \mathcal{M})$ with $\Gamma = \{\gamma_1, \gamma_2, \gamma_3\}$ and $\mathcal{M}\{\gamma_1\} = 0.5, \mathcal{M}\{\gamma_2\} = 0.4, \mathcal{M}\{\gamma_3\} = 0.3$. The uncertain variable ξ defined by

$$\xi(\gamma) = \begin{cases} -1, & \text{if } \gamma = \gamma_1 \\ 0, & \text{if } \gamma = \gamma_2 \\ 1, & \text{if } \gamma = \gamma_3 \end{cases}$$

has an uncertainty distribution

$$\Phi(x) = \begin{cases} 0, & \text{if } x < -1 \\ 0.5, & \text{if } -1 \leq x < 0 \\ 0.7, & \text{if } 0 \leq x < 1 \\ 1, & \text{if } x \geq 1. \end{cases}$$

Uncertain variables are said to be identically distributed if they share a common uncertainty distribution. Peng and Iwamura [57] showed that a function $\Phi : \Re \to [0, 1]$ is an uncertainty distribution if and only if it is a monotone increasing function except $\Phi(x) \equiv 0$ and $\Phi(x) \equiv 1$. Interested readers may refer to Peng and Iwamura [57] or Liu [40] for details.

Example 2.7 An uncertain variable ξ is called linear if it has a linear uncertainty distribution

$$\Phi(x) = \begin{cases} 0, & \text{if } x \leq a \\ (x - a)/(b - a), & \text{if } a \leq x \leq b \\ 1, & \text{if } x \geq b \end{cases}$$

where a and b are real numbers with $a < b$. For simplicity, this could be denoted by $\xi \sim \mathcal{L}(a, b)$.

Example 2.8 An uncertain variable ξ is called normal if it has a normal uncertainty distribution

$$\Phi(x) = \left(1 + \exp\left(\frac{\pi(e - x)}{\sqrt{3}\sigma}\right)\right)^{-1}, \quad x \in \Re$$

where e and σ are real numbers with $\sigma > 0$. For simplicity, this could be denoted by $\xi \sim \mathcal{N}(e, \sigma)$.

Example 2.9 An uncertain variable ξ is called lognormal if it has a lognormal uncertainty distribution

$$\Phi(x) = \left(1 + \exp\left(\frac{\pi(e - \ln x)}{\sqrt{3}\sigma}\right)\right)^{-1}, \quad x \geq 0$$

where e and σ are real numbers with $\sigma > 0$. For simplicity, this could be denoted by $\xi \sim \mathcal{LOGN}(e, \sigma)$.

Operational Law

Theorem 2.5 (Liu [40]) *Let $\xi_1, \xi_2, \ldots, \xi_n$ be independent uncertain variables with continuous uncertainty distributions $\Phi_1, \Phi_2, \ldots, \Phi_n$, respectively. If the function $f(x_1, x_2, \ldots, x_n)$ is strictly increasing with respect to $x_1, x_2, \ldots, x_m$ and strictly decreasing with respect to $x_{m+1}, x_{m+2}, \ldots, x_n$, then the uncertain variable*

$$\xi = f(\xi_1, \xi_2, \ldots, \xi_n) \tag{2.11}$$

has an uncertainty distribution

$$\Phi(x) = \sup_{f(x_1, x_2, \ldots, x_n) \leq x} \left(\min_{1 \leq i \leq m} \Phi_i(x_i) \wedge \min_{m+1 \leq i \leq n} (1 - \Phi_i(x_i)) \right). \tag{2.12}$$

Proof For simplicity, we only prove the case of $m = 1$ and $n = 2$. Note that $f(x_1, x_2)$ is strictly increasing with respect to x_1 and strictly decreasing with respect to x_2, and ξ_1 and ξ_2 are independent uncertain variables. On the one hand, we have

$$\mathcal{M}\{f(\xi_1, \xi_2) \leq x\} = \mathcal{M}\left\{ \bigcup_{f(x_1, x_2) \leq x} (\xi_1 \leq x_1) \cap (\xi_2 \geq x_2) \right\}$$

$$\geq \sup_{f(x_1, x_2) \leq x} \mathcal{M}\{(\xi_1 \leq x_1) \cap (\xi_2 \geq x_2)\} = \sup_{f(x_1, x_2) \leq x} \Phi_1(x_1) \wedge (1 - \Phi_2(x_2)).$$

On the other hand, we have

$$\mathcal{M}\{f(\xi_1, \xi_2) > x\} = \mathcal{M}\left\{ \bigcup_{f(x_1,x_2)>x} (\xi_1 \geq x_1) \cap (\xi_2 \leq x_2) \right\}$$

$$\geq \sup_{f(x_1,x_2)>x} \mathcal{M}\{(\xi_1 \geq x_1) \cap (\xi_2 \leq x_2)\} = \sup_{f(x_1,x_2)>x} (1 - \Phi_1(x_1)) \wedge \Phi_2(x_2).$$

Since

$$\sup_{f(x_1,x_2)\leq x} \Phi_1(x_1) \wedge (1 - \Phi_2(x_2)) + \sup_{f(x_1,x_2)>x} (1 - \Phi_1(x_1)) \wedge \Phi_2(x_2) = 1$$

and

$$\mathcal{M}\{f(\xi_1, \xi_2) \leq x\} + \mathcal{M}\{f(\xi_1, \xi_2) > x\} = 1,$$

we have

$$\Phi(x) = \mathcal{M}\{f(\xi_1, \xi_2) \leq x\} = \sup_{f(x_1,x_2)\leq x} \Phi_1(x_1) \wedge (1 - \Phi_2(x_2)).$$

The theorem is proved.

Remark 2.1 If f is a strictly increasing function, then $\xi = f(\xi_1, \xi_2, \ldots, \xi_n)$ has an uncertainty distribution

$$\Phi(x) = \sup_{f(x_1,x_2,\ldots,x_n)\leq x} \min_{1\leq i\leq n} \Phi_i(x_i).$$

Remark 2.2 If f is a strictly decreasing function, then $\xi = f(\xi_1, \xi_2, \ldots, \xi_n)$ has an uncertainty distribution

$$\Phi(x) = \sup_{f(x_1,x_2,\ldots,x_n)\leq x} \min_{1\leq i\leq n} (1 - \Phi_i(x_i)).$$

Example 2.10 Assume that ξ_1 and ξ_2 are independent uncertain variables with continuous uncertainty distributions Φ_1 and Φ_2, respectively. Then $\xi_1 \vee \xi_2$ has an uncertainty distribution

$$\Psi(x) = \sup_{x_1 \vee x_2 \leq x} \Phi_1(x_1) \wedge \Phi_2(x_2) = \Phi_1(x) \wedge \Phi_2(x),$$

and $\xi_1 \wedge \xi_2$ has an uncertainty distribution

$$\Upsilon(x) = \sup_{x_1 \wedge x_2 \leq x} \Phi_1(x_1) \wedge \Phi_2(x_2) = \Phi_1(x) \vee \Phi_2(x).$$

Example 2.11 Assume that ξ_1 and ξ_2 are independent uncertain variables with continuous uncertainty distributions Φ_1 and Φ_2, respectively. Then $\xi_1 + \xi_2$ has an uncertainty distribution

$$\Psi(x) = \sup_{x_1+x_2 \leq x} \Phi_1(x_1) \wedge \Phi_2(x_2) = \sup_{y \in \Re} \Phi_1(x - y) \wedge \Phi_2(y),$$

and $\xi_1 - \xi_2$ has an uncertainty distribution

$$\Upsilon(x) = \sup_{x_1-x_2 \leq x} \Phi_1(x_1) \wedge (1 - \Phi_2(x_2)) = \sup_{y \in \Re} \Phi_1(x + y) \wedge (1 - \Phi_2(y)).$$

Example 2.12 Assume that ξ_1 and ξ_2 are independent and positive uncertain variables with continuous uncertainty distributions Φ_1 and Φ_2, respectively. Then $\xi_1 \cdot \xi_2$ has an uncertainty distribution

$$\Psi(x) = \sup_{x_1 \cdot x_2 \leq x} \Phi_1(x_1) \wedge \Phi_2(x_2) = \sup_{y > 0} \Phi_1(x/y) \wedge \Phi_2(y),$$

and ξ_1/ξ_2 has an uncertainty distribution

$$\Upsilon(x) = \sup_{x_1/x_2 \leq x} \Phi_1(x_1) \wedge (1 - \Phi_2(x_2)) = \sup_{y > 0} \Phi_1(xy) \wedge (1 - \Phi_2(y)).$$

2.4 Inverse Uncertainty Distribution

Definition 2.7 (*Liu [40]*) An uncertainty distribution Φ is called regular if it is a continuous and strictly increasing function, and

$$\lim_{x \to -\infty} \Phi(x) = 0, \quad \lim_{x \to +\infty} \Phi(x) = 1.$$

Note that a regular uncertainty distribution Φ has an inverse function Φ^{-1} on the open interval $(0, 1)$. Besides, the domain of Φ^{-1} could be extended to $[0, 1]$ via

$$\Phi^{-1}(0) = \lim_{\alpha \downarrow 0} \Phi^{-1}(\alpha), \quad \Phi^{-1}(1) = \lim_{\alpha \uparrow 1} \Phi^{-1}(\alpha)$$

provided that the limits exist.

Definition 2.8 (*Liu [40]*) Let ξ be an uncertain variable with a regular uncertainty distribution Φ. Then the inverse function Φ^{-1} is called the inverse uncertainty distribution of ξ.

Liu [43] showed that a function $\Phi^{-1} : (0, 1) \to \Re$ is an inverse uncertainty distribution if and only if it is a continuous and strictly increasing function.

Example 2.13 The inverse uncertainty distribution of a linear uncertain variable
$\mathcal{L}(a, b)$ is

$$\Phi^{-1}(\alpha) = (1 - \alpha)a + \alpha b.$$

Example 2.14 The inverse uncertainty distribution of a normal uncertain variable
$\mathcal{N}(e, \sigma)$ is

$$\Phi^{-1}(\alpha) = e + \frac{\sqrt{3}\sigma}{\pi} \ln \frac{\alpha}{1 - \alpha}.$$

Example 2.15 The inverse uncertainty distribution of a lognormal uncertain variable
$\mathcal{LOGN}(e, \sigma)$ is

$$\Phi^{-1}(\alpha) = \exp(e) \left(\frac{\alpha}{1 - \alpha} \right)^{\sqrt{3}\sigma/\pi}.$$

Operational Law

Theorem 2.6 (Liu [40]) *Let $\xi_1, \xi_2, \ldots, \xi_n$ be independent uncertain variables
with regular uncertainty distributions $\Phi_1, \Phi_2, \ldots, \Phi_n$, respectively. If the function
$f(x_1, x_2, \ldots, x_n)$ is strictly increasing with respect to $x_1, x_2, \ldots, x_m$ and strictly
decreasing with respect to $x_{m+1}, x_{m+2}, \ldots, x_n$, then the uncertain variable*

$$\xi = f(\xi_1, \xi_2, \ldots, \xi_n) \tag{2.13}$$

has an inverse uncertainty distribution

$$\Phi^{-1}(\alpha) = f(\Phi_1^{-1}(\alpha), \ldots, \Phi_m^{-1}(\alpha), \Phi_{m+1}^{-1}(1 - \alpha), \ldots, \Phi_n^{-1}(1 - \alpha)). \tag{2.14}$$

Proof For simplicity, we only prove the case of $m = 1$ and $n = 2$. Note that $f(x_1, x_2)$
is strictly increasing with respect to x_1 and strictly decreasing with respect to x_2, and
ξ_1 and ξ_2 are independent uncertain variables. On the one hand, we have

$$\{f(\xi_1, \xi_2) \le f(\Phi_1^{-1}(\alpha), \Phi_2^{-1}(1 - \alpha))\} \supset \{\xi_1 \le \Phi_1^{-1}(\alpha)\} \cap \{\xi_2 \ge \Phi_2^{-1}(1 - \alpha)\}$$

and

$$\mathcal{M}\{\xi \le \Phi^{-1}(\alpha)\} \ge \mathcal{M}\{\xi_1 \le \Phi_1^{-1}(\alpha)\} \wedge \mathcal{M}\{\xi_2 \ge \Phi_2^{-1}(1 - \alpha)\} = \alpha \wedge \alpha = \alpha.$$

On the other hand, we have

$$\{f(\xi_1, \xi_2) \le f(\Phi_1^{-1}(\alpha), \Phi_2^{-1}(1 - \alpha))\} \subset \{\xi_1 \le \Phi_1^{-1}(\alpha)\} \cup \{\xi_2 \ge \Phi_2^{-1}(1 - \alpha)\}$$

and

$$\mathcal{M}\{\xi \le \Phi^{-1}(\alpha)\} \le \mathcal{M}\{\xi_1 \le \Phi_1^{-1}(\alpha)\} \vee \mathcal{M}\{\xi_2 \ge \Phi_2^{-1}(1 - \alpha)\} = \alpha \vee \alpha = \alpha.$$

Hence we have $\mathcal{M}\{\xi \leq \Phi^{-1}(\alpha)\} = \alpha$, which implies Φ is just the uncertainty distribution of ξ. The theorem is proved.

Remark 2.3 If f is a strictly increasing function, then $\xi = f(\xi_1, \xi_2, \ldots, \xi_n)$ has an inverse uncertainty distribution

$$\Phi^{-1}(\alpha) = f(\Phi_1^{-1}(\alpha), \Phi_2^{-1}(\alpha), \ldots, \Phi_n^{-1}(\alpha)).$$

Remark 2.4 If f is a strictly decreasing function, then $\xi = f(\xi_1, \xi_2, \ldots, \xi_n)$ has an inverse uncertainty distribution

$$\Phi^{-1}(\alpha) = f(\Phi_1^{-1}(1 - \alpha), \Phi_2^{-1}(1 - \alpha), \ldots, \Phi_n^{-1}(1 - \alpha)).$$

Example 2.16 Assume that ξ_1 and ξ_2 are independent uncertain variables with regular uncertainty distributions Φ_1 and Φ_2, respectively. Then $\xi_1 \vee \xi_2$ has an inverse uncertainty distribution

$$\Psi^{-1}(\alpha) = \Phi_1^{-1}(\alpha) \vee \Phi_2^{-1}(\alpha),$$

and $\xi_1 \wedge \xi_2$ has an inverse uncertainty distribution

$$\Upsilon^{-1}(\alpha) = \Phi_1^{-1}(\alpha) \wedge \Phi_2^{-1}(\alpha).$$

Example 2.17 Assume that ξ_1 and ξ_2 are independent uncertain variables with regular uncertainty distributions Φ_1 and Φ_2, respectively. Then $\xi_1 + \xi_2$ has an inverse uncertainty distribution

$$\Psi^{-1}(\alpha) = \Phi_1^{-1}(\alpha) + \Phi_2^{-1}(\alpha),$$

and $\xi_1 - \xi_2$ has an inverse uncertainty distribution

$$\Upsilon^{-1}(\alpha) = \Phi_1^{-1}(\alpha) - \Phi_2^{-1}(1 - \alpha).$$

Example 2.18 Assume that ξ_1 and ξ_2 are independent and positive uncertain variables with regular uncertainty distributions Φ_1 and Φ_2, respectively. Then $\xi_1 \cdot \xi_2$ has an inverse uncertainty distribution

$$\Psi^{-1}(\alpha) = \Phi_1^{-1}(\alpha) \cdot \Phi_2^{-1}(\alpha),$$

and ξ_1 / ξ_2 has an inverse uncertainty distribution

$$\Upsilon^{-1}(\alpha) = \Phi_1^{-1}(\alpha) / \Phi_2^{-1}(1 - \alpha).$$

2.5 Expected Value

Definition 2.9 (*Liu [36]*) Let ξ be an uncertain variable. Then its expected value $E[\xi]$ is defined by

$$E[\xi] = \int_0^{+\infty} \mathcal{M}\{\xi \geq x\}dx - \int_{-\infty}^0 \mathcal{M}\{\xi \leq x\}dx \qquad (2.15)$$

provided that at least one of the two integrals is finite.

Theorem 2.7 (Liu [36]) *Let ξ be an uncertain variable with an uncertainty distribution Φ. If the expected value $E[\xi]$ exists, then*

$$E[\xi] = \int_0^{+\infty} (1 - \Phi(x))dx - \int_{-\infty}^0 \Phi(x)dx. \qquad (2.16)$$

Proof Note that $\mathcal{M}\{\xi \geq x\} = 1 - \Phi(x)$ holds for almost every real number x and $\mathcal{M}\{\xi \leq x\}$ is just $\Phi(x)$. It follows from Definition 2.9 of expected value that

$$E[\xi] = \int_0^{+\infty} \mathcal{M}\{\xi \geq x\}dx - \int_{-\infty}^0 \mathcal{M}\{\xi \leq x\}dx$$

$$= \int_0^{+\infty} (1 - \Phi(x))dx - \int_{-\infty}^0 \Phi(x)dx.$$

The theorem is proved.

Theorem 2.8 (Liu [40]) *Let ξ be an uncertain variable with a regular uncertainty distribution Φ. If the expected value $E[\xi]$ exists, then*

$$E[\xi] = \int_0^1 \Phi^{-1}(\alpha)d\alpha. \qquad (2.17)$$

Proof By using the method of changing variables, it follows from Theorem 2.7 that

$$E[\xi] = \int_0^{+\infty} (1 - \Phi(x))dx - \int_{-\infty}^0 \Phi(x)dx$$

$$= \int_{\Phi(0)}^1 \Phi^{-1}(\alpha)d\alpha + \int_0^{\Phi(0)} \Phi^{-1}(\alpha)d\alpha = \int_0^1 \Phi^{-1}(\alpha)d\alpha.$$

The theorem is proved.

Example 2.19 The linear uncertain variable $\xi \sim \mathcal{L}(a, b)$ has an expected value

$$E[\xi] = \frac{a+b}{2}.$$

Example 2.20 The normal uncertain variable $\xi \sim \mathcal{N}(e, \sigma)$ has an expected value

$$E[\xi] = e.$$

Example 2.21 The lognormal uncertain variable $\xi \sim \mathcal{LOGN}(e, \sigma)$ has an expected value

$$E[\xi] = \begin{cases} \sqrt{3}\sigma \exp(e) \csc(\sqrt{3}\sigma), & \text{if } \sigma < \pi/\sqrt{3} \\ +\infty, & \text{if } \sigma \geq \pi/\sqrt{3}. \end{cases}$$

Interested readers may refer to Yao [71] for a detailed proof.

Theorem 2.9 (Liu and Ha [51]) *Let $\xi_1, \xi_2, \ldots, \xi_n$ be independent uncertain variables with regular uncertainty distributions $\Phi_1, \Phi_2, \ldots, \Phi_n$, respectively. If the function $f(x_1, x_2, \ldots, x_n)$ is strictly increasing with respect to $x_1, x_2, \ldots, x_m$ and strictly decreasing with respect to $x_{m+1}, x_{m+2}, \ldots, x_n$, then the uncertain variable*

$$\xi = f(\xi_1, \xi_2, \ldots, \xi_n) \tag{2.18}$$

has an expected value

$$E[\xi] = \int_0^1 f(\Phi_1^{-1}(\alpha), \ldots, \Phi_m^{-1}(\alpha), \Phi_{m+1}^{-1}(1-\alpha), \ldots, \Phi_n^{-1}(1-\alpha))d\alpha \tag{2.19}$$

provided that $E[\xi]$ exists.

Proof The theorem follows immediately from Theorems 2.6 and 2.8.

Example 2.22 Let ξ and η be independent and positive uncertain variables with regular uncertainty distributions Φ and Ψ, respectively. Then

$$E[\xi\eta] = \int_0^1 \Phi^{-1}(\alpha)\Psi^{-1}(\alpha)d\alpha,$$

and

$$E\left[\frac{\xi}{\eta}\right] = \int_0^1 \frac{\Phi^{-1}(\alpha)}{\Psi^{-1}(1-\alpha)}d\alpha.$$

Theorem 2.10 (Liu [36], Markov Inequality) *Let ξ be an uncertain variable. Then for any given real number $r > 0$, we have*

$$\mathcal{M}\{|\xi| \geq r\} \leq \frac{E[|\xi|]}{r}. \tag{2.20}$$

Proof It follows from Definition 2.9 of expected value that

$$E[|\xi|] = \int_0^{+\infty} \mathcal{M}\{|\xi| \geq x\}dx \geq \int_0^r \mathcal{M}\{|\xi| \geq x\}dx$$

$$\geq \int_0^r \mathcal{M}\{|\xi| \geq r\}dx = r \cdot \mathcal{M}\{|\xi| \geq r\}.$$

Then we have

$$\mathcal{M}\{|\xi| \geq r\} \leq \frac{E[|\xi|]}{r}.$$

The theorem is proved.

2.6 Variance

Definition 2.10 *(Liu [36])* Let ξ be an uncertain variable with a finite expected value e. Then its variance $V[\xi]$ is defined by

$$V[\xi] = E[(\xi - e)^2]. \tag{2.21}$$

Remark 2.5 Note that the variance of an uncertain variable is defined based on the expected value. If the expected value of an uncertain variable is infinite or does not exist at all, then the variance of the uncertain variable does not exist.

For an uncertain variable ξ with a finite expected value e, if we know only its uncertainty distribution Φ, then we can derive an upper bound of its variance $V[\xi]$ as follows:

$$V[\xi] = \int_0^{+\infty} \mathcal{M}\{(\xi - e)^2 \geq x\}dx$$

$$= \int_0^{+\infty} \mathcal{M}\{(\xi \geq e + \sqrt{x}) \cup (\xi \leq e - \sqrt{x})\}dx$$

$$\leq \int_0^{+\infty} (\mathcal{M}\{\xi \geq e + \sqrt{x}\} + \mathcal{M}\{\xi \leq e - \sqrt{x}\})dx$$

$$= \int_0^{+\infty} (1 - \Phi(e + \sqrt{x}) + \Phi(e - \sqrt{x}))dx.$$

In this case, we stipulate that

$$V[\xi] = \int_0^{+\infty} (1 - \Phi(e + \sqrt{x}) + \Phi(e - \sqrt{x}))dx. \tag{2.22}$$

Theorem 2.11 (Yao [84]) *Let ξ be an uncertain variable with a regular uncertainty distribution Φ. If its expected value e exists, then its variance*

$$V[\xi] = \int_0^1 (\Phi^{-1}(\alpha) - e)^2 d\alpha. \tag{2.23}$$

Proof It follows from Stipulation (2.22) that

$$V[\xi] = \int_0^{+\infty} (1 - \Phi(e + \sqrt{x}))dx + \int_0^{+\infty} \Phi(e - \sqrt{x})dx.$$

For the first term, substituting $\Phi(e + \sqrt{x})$ with α, and x with $\left(\Phi^{-1}(\alpha) - e\right)^2$ accordingly, we have

$$\int_0^{+\infty} (1 - \Phi(e + \sqrt{x}))dx$$

$$= \int_{\Phi(e)}^1 (1 - \alpha)d\left(\Phi^{-1}(\alpha) - e\right)^2$$

$$= (1 - \alpha)\left(\Phi^{-1}(\alpha) - e\right)^2\Big|_{\Phi(e)}^1 - \int_{\Phi(e)}^1 \left(\Phi^{-1}(\alpha) - e\right)^2 d(1 - \alpha)$$

$$= \int_{\Phi(e)}^1 \left(\Phi^{-1}(\alpha) - e\right)^2 d\alpha.$$

For the second term, substituting $\Phi(e - \sqrt{x})$ with α, and x with $\left(e - \Phi^{-1}(\alpha)\right)^2$ accordingly, we have

$$\int_0^{+\infty} \Phi(e - \sqrt{x})dx$$

$$= \int_{\Phi(e)}^0 \alpha d\left(e - \Phi^{-1}(\alpha)\right)^2$$

$$= \alpha\left(e - \Phi^{-1}(\alpha)\right)^2\Big|_{\Phi(e)}^0 - \int_{\Phi(e)}^0 \left(e - \Phi^{-1}(\alpha)\right)^2 d\alpha$$

$$= \int_0^{\Phi(e)} \left(\Phi^{-1}(\alpha) - e\right)^2 d\alpha.$$

Hence

$$V[\xi] = \int_{\Phi(e)}^1 \left(\Phi^{-1}(\alpha) - e\right)^2 d\alpha + \int_0^{\Phi(e)} \left(\Phi^{-1}(\alpha) - e\right)^2 d\alpha$$

$$= \int_0^1 \left(\Phi^{-1}(\alpha) - e\right)^2 d\alpha.$$

The theorem is proved.

Example 2.23 The linear uncertain variable $\xi \sim \mathcal{L}(a, b)$ has a variance

$$V[\xi] = \frac{(b-a)^2}{12}.$$

Example 2.24 The normal uncertain variable $\xi \sim \mathcal{N}(e, \sigma)$ has a variance

$$V[\xi] = \sigma^2.$$

Chapter 3
Uncertain Process

Uncertain process is a sequence of uncertain variables indexed by the time. The emphases of this chapter are on the concepts about an uncertain process including its uncertainty distribution, extreme value, and time integral.

3.1 Uncertain Process

Definition 3.1 (*Liu [37]*) Let T be an index set and $(\Gamma, \mathcal{L}, \mathcal{M})$ be an uncertainty space. An uncertain process X_t is a measurable function from $T \times (\Gamma, \mathcal{L}, \mathcal{M})$ to the set of real numbers, i.e., the set

$$\{X_t \in B\} = \{\gamma \in \Gamma \mid X_t(\gamma) \in B\} \tag{3.1}$$

is an event for each $t \in T$ and each Borel set B of real numbers.

Example 3.1 Let $\xi_1, \xi_2, \ldots$ be a sequence of uncertain variables. Then

$$X_n = \xi_1 + \xi_2 + \cdots + \xi_n$$

is an uncertain process.

An uncertain process $X_t(\gamma)$ is essentially a function of two variables t and γ. From the one aspect, for each time t^*, the function $X_{t^*}(\gamma)$ is an uncertain variable. Thus an uncertain process could be regarded as a sequence of uncertain variables indexed by the time. From the other aspect, for each γ^*, the function $X_t(\gamma^*)$ is a function of the time t which is called a sample path. Thus an uncertain process could be regarded as a mapping from an uncertainty space to the family of the sample paths.

Definition 3.2 Let X_t be an uncertain process. Then for any two times t_1 and t_2 with $t_1 < t_2$, the difference $X_{t_2} - X_{t_1}$ is called an increment of X_t.

© Springer-Verlag Berlin Heidelberg 2016
K. Yao, *Uncertain Differential Equations*,
Springer Uncertainty Research, DOI 10.1007/978-3-662-52729-0_3

Definition 3.3 (*Liu [40]*) An uncertain process X_t is said to be sample-continuous if its almost all sample paths are continuous.

Definition 3.4 (*Liu [46]*) Uncertain processes $X_{1t}, X_{2t}, \ldots, X_{nt}$ are said to be independent if for any positive integer m and any times $t_1, t_2, \ldots, t_m$, the uncertain vectors

$$\xi_i = (X_{it_1}, X_{it_2}, \ldots, X_{it_m}), \quad i = 1, 2, \ldots, n \qquad (3.2)$$

are independent.

Uncertainty Distribution

Definition 3.5 (*Liu [46]*) An uncertain process X_t is said to have an uncertainty distribution $\Phi_t(x)$ if at each time t, the uncertain variable X_t has an uncertainty distribution $\Phi_t(x)$.

Theorem 3.1 (Liu [46]) *Let $X_{1t}, X_{2t}, \ldots, X_{nt}$ be independent uncertain processes with continuous uncertainty distributions $\Phi_{1t}, \Phi_{2t}, \ldots, \Phi_{nt}$, respectively. If the function $f(x_1, x_2, \ldots, x_n)$ is strictly increasing with respect to $x_1, x_2, \ldots, x_m$ and strictly decreasing with respect to $x_{m+1}, x_{m+2}, \ldots, x_n$, then the uncertain process*

$$X_t = f(X_{1t}, X_{2t}, \ldots, X_{nt}) \qquad (3.3)$$

has an uncertainty distribution

$$\Phi_t(x) = \sup_{f(x_1, x_2, \ldots, x_n) \le x} \left(\min_{1 \le i \le m} \Phi_{it}(x_i) \wedge \min_{m+1 \le i \le n} (1 - \Phi_{it}(x_i)) \right). \qquad (3.4)$$

Proof Note that at each time t, the uncertain variables $X_{1t}, X_{2t}, \ldots, X_{nt}$ are independent. Then the theorem follows immediately from Theorem 2.5.

Inverse Uncertainty Distribution

Definition 3.6 (*Liu [46]*) An uncertain process X_t is said to have an inverse uncertainty distribution $\Phi_t^{-1}(\alpha)$ if at each time t, the uncertain variable X_t has an inverse uncertainty distribution $\Phi_t^{-1}(\alpha)$.

Theorem 3.2 (Liu [46]) *Let $X_{1t}, X_{2t}, \ldots, X_{nt}$ be independent uncertain processes with inverse uncertainty distributions $\Phi_{1t}^{-1}, \Phi_{2t}^{-1}, \ldots, \Phi_{nt}^{-1}$, respectively. If the function $f(x_1, x_2, \ldots, x_n)$ is strictly increasing with respect to $x_1, x_2, \ldots, x_m$ and strictly decreasing with respect to $x_{m+1}, x_{m+2}, \ldots, x_n$, then the uncertain process*

$$X_t = f(X_{1t}, X_{2t}, \ldots, X_{nt}) \qquad (3.5)$$

has an inverse uncertainty distribution

$$\Phi_t^{-1}(\alpha) = f(\Phi_{1t}^{-1}(\alpha), \ldots, \Phi_{mt}^{-1}(\alpha), \Phi_{m+1,t}^{-1}(1 - \alpha), \ldots, \Phi_{nt}^{-1}(1 - \alpha)). \qquad (3.6)$$

Proof Note that at each time t, the uncertain variables $X_{1t}, X_{2t}, \ldots, X_{nt}$ are independent. Then the theorem follows immediately from Theorem 2.6.

Stationary Independent Increment Process

Definition 3.7 (*Liu [37]*) An uncertain process X_t is called a stationary increment process if for any given $t > 0$, the increments $X_{s+t} - X_s$ are identically distributed for all $s > 0$.

Definition 3.8 (*Liu [37]*) An uncertain process X_t is called an independent increment process if for any times $t_0, t_1, t_2, \ldots, t_k$ with $t_0 < t_1 < \cdots < t_k$, the uncertain variables

$$X_{t_0}, \ X_{t_1} - X_{t_0}, \ X_{t_2} - X_{t_1}, \ldots, X_{t_k} - X_{t_{k-1}} \tag{3.7}$$

are independent.

Definition 3.9 (*Liu [37]*) An uncertain process is called a stationary independent increment process if it is not only a stationary increment process but also an independent increment process.

Example 3.2 Let $\xi_1, \xi_2, \ldots$ be a sequence of iid uncertain variables. Then

$$X_n = \xi_1 + \xi_2 + \cdots + \xi_n \tag{3.8}$$

is a stationary independent increment process.

3.2 Extreme Value

Definition 3.10 Let X_t be an uncertain process on the uncertainty space $(\Gamma, \mathcal{L}, \mathcal{M})$. Then its supremum process

$$Y_t = \sup_{0 \le s \le t} X_s \tag{3.9}$$

is defined by

$$Y_t(\gamma) = \sup_{0 \le s \le t} X_s(\gamma), \quad \forall \gamma \in \Gamma. \tag{3.10}$$

Theorem 3.3 *Let X_t be an uncertain process with an uncertainty distribution Φ_t. Then its supremum process Y_t has an uncertainty distribution Ψ_t such that*

$$\Psi_t(x) \le \inf_{0 \le s \le t} \Phi_s(x), \quad \forall x \in \Re. \tag{3.11}$$

Proof For any given time t and for any $s \in [0, t]$, we have

$$\{Y_t \leq x\} = \left\{ \sup_{0 \leq u \leq t} X_u \leq x \right\} \subset \{X_s \leq x\}, \quad \forall x \in \mathfrak{R}.$$

It follows from the monotonicity of uncertain measure that

$$\Psi_t(x) = \mathcal{M}\{Y_t \leq x\} \leq \mathcal{M}\{X_s \leq x\} = \Phi_s(x), \quad \forall x \in \mathfrak{R}.$$

Since the above inequality holds for all $s \in [0, t]$, we have

$$\Psi_t(x) \leq \inf_{0 \leq s \leq t} \Phi_s(x), \quad \forall x \in \mathfrak{R}.$$

The theorem is proved.

Definition 3.11 Let X_t be an uncertain process on the uncertainty space $(\Gamma, \mathcal{L}, \mathcal{M})$. Then its infimum process

$$Z_t = \inf_{0 \leq s \leq t} X_s \tag{3.12}$$

is defined by

$$Z_t(\gamma) = \inf_{0 \leq s \leq t} X_s(\gamma), \quad \forall \gamma \in \Gamma. \tag{3.13}$$

Theorem 3.4 *Let X_t be an uncertain process with an uncertainty distribution Φ_t. Then its infimum process Z_t has an uncertainty distribution Υ_t such that*

$$\Upsilon_t(x) \geq \sup_{0 \leq s \leq t} \Phi_s(x), \quad \forall x \in \mathfrak{R}. \tag{3.14}$$

Proof For any given time t and for any $s \in [0, t]$, we have

$$\{Z_t \leq x\} = \left\{ \inf_{0 \leq u \leq t} X_u \leq x \right\} \supset \{X_s \leq x\}, \quad \forall x \in \mathfrak{R}.$$

It follows from the monotonicity of uncertain measure that

$$\Upsilon_t(x) = \mathcal{M}\{Z_t \leq x\} \geq \mathcal{M}\{X_s \leq x\} = \Phi_s(x), \quad \forall x \in \mathfrak{R}.$$

Since the above inequality holds for all $s \in [0, t]$, we have

$$\Upsilon_t(x) \geq \sup_{0 \leq s \leq t} \Phi_s(x), \quad \forall x \in \mathfrak{R}.$$

The theorem is proved.

3.3   Time Integral

Definition 3.12   (*Liu [37]*) Let X_t be an uncertain process. For any partition of closed interval $[a, b]$ with $a = t_1 < t_2 < \cdots < t_{k+1} = b$, the mesh is written as

$$\Delta = \max_{1 \le i \le k} |t_{i+1} - t_i|. \tag{3.15}$$

Then the time integral of X_t with respect to t on the interval $[a, b]$ is defined by

$$\int_a^b X_t dt = \lim_{\Delta \to 0} \sum_{i=1}^{k} X_{t_i} \cdot (t_{i+1} - t_i) \tag{3.16}$$

provided that the limit exists almost surely and is finite. In this case, the uncertain process X_t is said to be time integrable.

Remark 3.1   Assume that X_t is a time integrable uncertain process on an uncertainty space $(\Gamma, \mathcal{L}, \mathcal{M})$. Then for each $\gamma \in \Gamma$, the integral

$$\int_0^t X_s(\gamma) ds$$

is a sample path of

$$\int_0^t X_s ds.$$

Theorem 3.5   (Liu [47]) *If X_t is a sample-continuous uncertain process on $[a, b]$, then it is time integrable on $[a, b]$, and*

$$\left| \int_a^b X_t(\gamma) dt \right| \le \int_a^b |X_t(\gamma)| dt \tag{3.17}$$

for each sample path $X_t(\gamma)$.

Proof Since the uncertain process X_t is sample-continuous, its almost all sample paths are continuous functions of the time t. So the limit

$$\lim_{\Delta \to 0} \sum_{i=1}^{k} X_{t_i} (t_{i+1} - t_i)$$

exists almost surely and is finite for any partition of the closed interval $[a, b]$, and the uncertain process X_t is time integrable on $[a, b]$. Note that $X_t(\gamma) \le |X_t(\gamma)|$ and $-X_t(\gamma) \le |X_t(\gamma)|$ hold for each sample path $X_t(\gamma)$ and for each $t \in [a, b]$. Then we have

$$\int_a^b X_t(\gamma) dt \le \int_a^b |X_t(\gamma)| dt, \quad -\int_a^b X_t(\gamma) dt \le \int_a^b |X_t(\gamma)| dt,$$

which imply

$$\left| \int_a^b X_t(\gamma) dt \right| \le \int_a^b |X_t(\gamma)| dt.$$

The theorem is proved.

Theorem 3.6 (Liu [47]) *If X_t is a time integrable uncertain process on $[a, b]$, then it is time integrable on each subinterval of $[a, b]$. Moreover, if $c \in [a, b]$, then*

$$\int_a^b X_t dt = \int_a^c X_t dt + \int_c^b X_t dt. \tag{3.18}$$

Proof Let $[a', b']$ be a subinterval of $[a, b]$. Since X_t is a time integrable uncertain process on $[a, b]$, for any partition

$$a = t_1 < \cdots < t_m = a' < t_{m+1} < \cdots < t_n = b' < t_{n+1} < \cdots < t_{k+1} = b,$$

the limit

$$\lim_{\Delta \to 0} \sum_{i=1}^{k} X_{t_i} (t_{i+1} - t_i)$$

exists almost surely and is finite. So the limit

$$\lim_{\Delta \to 0} \sum_{i=m}^{n-1} X_{t_i} (t_{i+1} - t_i)$$

exists almost surely and is finite, and the uncertain process X_t is time integrable on $[a', b']$. Next, for any partition of the closed interval $[a, b]$ with $t_m = c$ for some integer m, it follows from Definition 3.12 of time integral that

$$\int_a^b X_t dt = \lim_{\Delta \to 0} \sum_{i=1}^{k} X_{t_i} (t_{i+1} - t_i)$$

$$= \lim_{\Delta \to 0} \sum_{i=1}^{m-1} X_{t_i} (t_{i+1} - t_i) + \lim_{\Delta \to 0} \sum_{i=m}^{k} X_{t_i} (t_{i+1} - t_i)$$

$$= \int_a^c X_t dt + \int_c^b X_t dt.$$

The theorem is proved.

Theorem 3.7 (Liu [47], Linearity of Time Integral) *Let X_t and Y_t be two time integrable uncertain processes on $[a, b]$. Then*

$$\int_a^b (\alpha X_t + \beta Y_t)\mathrm{d}t = \alpha \int_a^b X_t\mathrm{d}t + \beta \int_a^b Y_t\mathrm{d}t \qquad (3.19)$$

for any real numbers α and β.

Proof For any partition of the closed interval $[a, b]$, it follows from Definition 3.12 of time integral that

$$\int_a^b (\alpha X_t + \beta Y_t)\mathrm{d}t = \lim_{\Delta \to 0} \sum_{i=1}^{k} (\alpha X_{t_i} + \beta Y_{t_i})(t_{i+1} - t_i)$$

$$= \lim_{\Delta \to 0} \alpha \sum_{i=1}^{k} X_{t_i}(t_{i+1} - t_i) + \lim_{\Delta \to 0} \beta \sum_{i=1}^{k} Y_{t_i}(t_{i+1} - t_i)$$

$$= \alpha \int_a^b X_t\mathrm{d}t + \beta \int_a^b Y_t\mathrm{d}t.$$

The theorem is proved.

Chapter 4
Contour Process

Contour process is a type of uncertain processes with some special structures so that the set of contour processes is closed under the extreme value operator, time integral operator, and monotone function operator. This chapter introduces contour processes, including their inverse uncertainty distributions, extreme values, time integrals, and monotone functions.

4.1 Contour Process

Definition 4.1 (*Yao [85]*) Let X_t be an uncertain process. If for each $\alpha \in (0, 1)$, there exists a real function X_t^α such that

$$\mathcal{M}\{X_t \leq X_t^\alpha, \forall t\} = \alpha, \tag{4.1}$$

$$\mathcal{M}\{X_t > X_t^\alpha, \forall t\} = 1 - \alpha, \tag{4.2}$$

then X_t is called a contour process. In this case, X_t^α is called an α-path of the contour process X_t.

Example 4.1 Let $\xi_1, \xi_2, \ldots$ be a sequence of independent uncertain variables. Then the uncertain process

$$X_n = \xi_1 + \xi_2 + \ldots + \xi_n$$

is a contour process.

Example 4.2 Let ξ be an uncertain variable. Then the uncertain process

$$X_n = \begin{cases} \xi, & \text{if } n \text{ is odd} \\ -\xi, & \text{if } n \text{ is even} \end{cases}$$

is not a contour process.

© Springer-Verlag Berlin Heidelberg 2016
K. Yao, *Uncertain Differential Equations*,
Springer Uncertainty Research, DOI 10.1007/978-3-662-52729-0_4

Theorem 4.1 (Yao [85]) *An uncertain process X_t is a contour process if and only if for each $\alpha \in (0, 1)$, there exists a real function X_t^α such that*

$$\mathcal{M}\{X_t < X_t^\alpha, \forall t\} = \alpha, \tag{4.3}$$

$$\mathcal{M}\{X_t \geq X_t^\alpha, \forall t\} = 1 - \alpha. \tag{4.4}$$

Proof On the one hand, assume that X_t is a contour process with an α-path X_t^α. Note that for each $\alpha \in (0, 1)$ and for any given $\varepsilon > 0$, we have

$$\{X_t \leq X_t^{\alpha-\varepsilon}, \forall t\} \subset \{X_t < X_t^\alpha, \forall t\} \subset \{X_t \leq X_t^\alpha, \forall t\},$$

$$\{X_t > X_t^\alpha, \forall t\} \subset \{X_t \geq X_t^\alpha, \forall t\} \subset \{X_t > X_t^{\alpha-\varepsilon}, \forall t\}.$$

By using the monotonicity of uncertain measure, we get that

$$\mathcal{M}\{X_t \leq X_t^{\alpha-\varepsilon}, \forall t\} \leq \mathcal{M}\{X_t < X_t^\alpha, \forall t\} \leq \mathcal{M}\{X_t \leq X_t^\alpha, \forall t\},$$

$$\mathcal{M}\{X_t > X_t^\alpha, \forall t\} \leq \mathcal{M}\{X_t \geq X_t^\alpha, \forall t\} \leq \mathcal{M}\{X_t > X_t^{\alpha-\varepsilon}, \forall t\}.$$

According to Definition 4.1 of contour process, we have

$$\mathcal{M}\{X_t \leq X_t^{\alpha-\varepsilon}, \forall t\} = \alpha - \varepsilon, \quad \mathcal{M}\{X_t \leq X_t^\alpha, \forall t\} = \alpha,$$

$$\mathcal{M}\{X_t > X_t^\alpha, \forall t\} = 1 - \alpha, \quad \mathcal{M}\{X_t > X_t^{\alpha-\varepsilon}, \forall t\} = 1 - \alpha + \varepsilon.$$

Then Eqs. (4.3) and (4.4) follow immediately. On the other hand, assume that Eqs. (4.3) and (4.4) hold for each $\alpha \in (0, 1)$. Note that for each $\alpha \in (0, 1)$ and for any given $\varepsilon > 0$, we have

$$\{X_t < X_t^\alpha, \forall t\} \subset \{X_t \leq X_t^\alpha, \forall t\} \subset \{X_t < X_t^{\alpha+\varepsilon}, \forall t\},$$

$$\{X_t \geq X_t^{\alpha+\varepsilon}, \forall t\} \subset \{X_t > X_t^\alpha, \forall t\} \subset \{X_t \geq X_t^\alpha, \forall t\}.$$

By using the monotonicity of uncertain measure, we get that

$$\mathcal{M}\{X_t < X_t^\alpha, \forall t\} \leq \mathcal{M}\{X_t \leq X_t^\alpha, \forall t\} \leq \mathcal{M}\{X_t < X_t^{\alpha+\varepsilon}, \forall t\},$$

$$\mathcal{M}\{X_t \geq X_t^{\alpha+\varepsilon}, \forall t\} \leq \mathcal{M}\{X_t > X_t^\alpha, \forall t\} \leq \mathcal{M}\{X_t \geq X_t^\alpha, \forall t\}.$$

It follows from Eqs. (4.3) and (4.4) that

$$\mathcal{M}\{X_t < X_t^\alpha, \forall t\} = \alpha, \quad \mathcal{M}\{X_t < X_t^{\alpha+\varepsilon}, \forall t\} = \alpha + \varepsilon,$$

$$\mathcal{M}\{X_t \geq X_t^{\alpha+\varepsilon}, \forall t\} = 1 - \alpha - \varepsilon, \quad \mathcal{M}\{X_t \geq X_t^\alpha, \forall t\} = 1 - \alpha.$$

Then

$$\mathcal{M}\{X_t \leq X_t^\alpha, \forall t\} = \alpha, \quad \mathcal{M}\{X_t > X_t^\alpha, \forall t\} = 1 - \alpha,$$

and the uncertain process X_t is a contour process. The theorem is proved.

4.2 Inverse Uncertainty Distribution

The α-path X_t^α of a contour process X_t is just its inverse uncertainty distribution.

Theorem 4.2 (Yao [85]) *Let X_t be a contour process with an α-path X_t^α. Then X_t has an inverse uncertainty distribution*

$$\Phi_t^{-1}(\alpha) = X_t^\alpha, \quad \forall \alpha \in (0, 1).$$

Proof Given any time s, since

$$\{X_s \leq X_s^\alpha\} \supset \{X_t \leq X_t^\alpha, \forall t\},$$

$$\{X_s > X_s^\alpha\} \supset \{X_t > X_t^\alpha, \forall t\},$$

we have

$$\mathcal{M}\{X_s \leq X_s^\alpha\} \geq \mathcal{M}\{X_t \leq X_t^\alpha, \forall t\} = \alpha,$$

$$\mathcal{M}\{X_s > X_s^\alpha\} \geq \mathcal{M}\{X_t > X_t^\alpha, \forall t\} = 1 - \alpha$$

according to the monotonicity of uncertain measure. It follows from the duality of uncertain measure that

$$\mathcal{M}\{X_s \leq X_s^\alpha\} + \mathcal{M}\{X_s > X_s^\alpha\} = 1.$$

Then we have

$$\mathcal{M}\{X_s \leq X_s^\alpha\} = \alpha, \quad \forall s$$

which means X_t has an inverse uncertainty distribution X_t^α. The theorem is proved.

Theorem 4.3 (Yao [85]) *Let X_t be a contour process with an α-path X_t^α. Then*

$$E[X_t] = \int_0^1 X_t^\alpha d\alpha.$$

Proof The theorem follows immediately from Theorems 2.8 and 4.2.

4.3   Extreme Value

The set of contour processes is closed under the supremum operator and the infimum operator.

Theorem 4.4 (Yao [85]) *Let X_t be a contour process with an α-path X_t^α. Then its supremum process*

$$Y_t = \sup_{0 \le s \le t} X_s$$

is a contour process with an α-path

$$Y_t^\alpha = \sup_{0 \le s \le t} X_s^\alpha.$$

Proof For a sample path $X_t(\gamma)$ such that $X_t(\gamma) \le X_t^\alpha$ for any time t, we have

$$\sup_{0 \le s \le t} X_s(\gamma) \le \sup_{0 \le s \le t} X_s^\alpha, \quad \forall t.$$

It implies

$$\left\{ Y_t \le \sup_{0 \le s \le t} X_s^\alpha, \forall t \right\} = \left\{ \sup_{0 \le s \le t} X_s \le \sup_{0 \le s \le t} X_s^\alpha, \forall t \right\} \supset \left\{ X_t \le X_t^\alpha, \forall t \right\}.$$

By the monotonicity of uncertain measure, we have

$$\mathcal{M} \left\{ Y_t \le \sup_{0 \le s \le t} X_s^\alpha, \forall t \right\} \ge \mathcal{M} \left\{ X_t \le X_t^\alpha, \forall t \right\} = \alpha.$$

Similarly, we have

$$\mathcal{M} \left\{ Y_t > \sup_{0 \le s \le t} X_s^\alpha, \forall t \right\} \ge \mathcal{M} \left\{ X_t > X_t^\alpha, \forall t \right\} = 1 - \alpha.$$

Since

$$\mathcal{M} \left\{ Y_t \le \sup_{0 \le s \le t} X_s^\alpha, \forall t \right\} + \mathcal{M} \left\{ Y_t > \sup_{0 \le s \le t} X_s^\alpha, \forall t \right\} \le 1,$$

we have

$$\mathcal{M} \left\{ Y_t \le \sup_{0 \le s \le t} X_s^\alpha, \forall t \right\} = \alpha,$$

$$\mathcal{M} \left\{ Y_t > \sup_{0 \le s \le t} X_s^\alpha, \forall t \right\} = 1 - \alpha.$$

So Y_t is a contour process with an α-path

$$Y_t^\alpha = \sup_{0 \le s \le t} X_s^\alpha.$$

The theorem is proved.

Theorem 4.5 (Yao [85]) *Let X_t be a contour process with an α-path X_t^α. Then its infimum process*

$$Z_t = \inf_{0 \le s \le t} X_s$$

is a contour process with an α-path

$$Z_t^\alpha = \inf_{0 \le s \le t} X_s^\alpha.$$

Proof For a sample path $X_t(\gamma)$ such that $X_t(\gamma) \le X_t^\alpha$ for any time t, we have

$$\inf_{0 \le s \le t} X_s(\gamma) \le \inf_{0 \le s \le t} X_s^\alpha, \quad \forall t.$$

It implies

$$\left\{ Z_t \le \inf_{0 \le s \le t} X_s^\alpha, \forall t \right\} = \left\{ \inf_{0 \le s \le t} X_s \le \inf_{0 \le s \le t} X_s^\alpha, \forall t \right\} \supset \left\{ X_t \le X_t^\alpha, \forall t \right\}.$$

By the monotonicity of uncertain measure, we have

$$\mathcal{M} \left\{ Z_t \le \inf_{0 \le s \le t} X_s^\alpha, \forall t \right\} \ge \mathcal{M} \left\{ X_t \le X_t^\alpha, \forall t \right\} = \alpha.$$

Similarly, we have

$$\mathcal{M} \left\{ Z_t > \inf_{0 \le s \le t} X_s^\alpha, \forall t \right\} \ge \mathcal{M} \left\{ X_t > X_t^\alpha, \forall t \right\} = 1 - \alpha.$$

Since

$$\mathcal{M} \left\{ Z_t \le \inf_{0 \le s \le t} X_s^\alpha, \forall t \right\} + \mathcal{M} \left\{ Z_t > \inf_{0 \le s \le t} X_s^\alpha, \forall t \right\} \le 1,$$

we have

$$\mathcal{M} \left\{ Z_t \le \inf_{0 \le s \le t} X_s^\alpha, \forall t \right\} = \alpha,$$

$$\mathcal{M} \left\{ Z_t > \inf_{0 \le s \le t} X_s^\alpha, \forall t \right\} = 1 - \alpha.$$

So Z_t is a contour process with an α-path

$$Z_t^\alpha = \inf_{0 \le s \le t} X_s^\alpha.$$

The theorem is proved.

4.4 Time Integral

The set of contour processes is closed under the time integral operator.

Theorem 4.6 (Yao [85]) *Let X_t be a contour process with an α-path X_t^α. Then its time integral process*

$$Y_t = \int_0^t X_s \, \mathrm{d}s$$

is a contour process with an α-path

$$Y_t^\alpha = \int_0^t X_s^\alpha \, \mathrm{d}s.$$

Proof For a sample path $X_t(\gamma)$ such that $X_t(\gamma) \le X_t^\alpha$ for any time t, we have

$$\int_0^t X_s(\gamma) \mathrm{d}s \le \int_0^t X_s^\alpha \mathrm{d}s, \quad \forall t.$$

It implies

$$\left\{ Y_t \le \int_0^t X_s^\alpha \mathrm{d}s, \forall t \right\} = \left\{ \int_0^t X_s \mathrm{d}s \le \int_0^t X_s^\alpha \mathrm{d}s, \forall t \right\} \supset \left\{ X_t \le X_t^\alpha, \forall t \right\}.$$

By the monotonicity of uncertain measure, we have

$$\mathcal{M} \left\{ Y_t \le \int_0^t X_s^\alpha \mathrm{d}s, \forall t \right\} \ge \mathcal{M} \left\{ X_t \le X_t^\alpha, \forall t \right\} = \alpha.$$

Similarly, we have

$$\mathcal{M} \left\{ Y_t > \int_0^t X_s^\alpha \mathrm{d}s, \forall t \right\} \ge \mathcal{M} \left\{ X_t > X_t^\alpha, \forall t \right\} = 1 - \alpha.$$

Since

$$\mathcal{M} \left\{ Y_t \le \int_0^t X_s^\alpha \mathrm{d}s, \forall t \right\} + \mathcal{M} \left\{ Y_t > \int_0^t X_s^\alpha \mathrm{d}s, \forall t \right\} \le 1,$$

we have

$$\mathcal{M}\left\{Y_t \leq \int_0^t X_s^\alpha ds, \forall t\right\} = \alpha,$$

$$\mathcal{M}\left\{Y_t > \int_0^t X_s^\alpha ds, \forall t\right\} = 1 - \alpha.$$

So Y_t is a contour process with an α-path

$$Y_t^\alpha = \int_0^t X_s^\alpha ds.$$

The theorem is proved.

4.5 Monotone Function

The set of contour processes is closed under the operators of monotone functions.

Theorem 4.7 (Yao [85]) *Let $X_{1t}, X_{2t}, \ldots, X_{nt}$ be independent contour processes with α-paths $X_{1t}^\alpha, X_{2t}^\alpha, \ldots, X_{nt}^\alpha$, respectively. If the function $f(x_1, x_2, \ldots, x_n)$ is strictly increasing with respect to $x_1, x_2, \ldots, x_m$ and strictly decreasing with respect to $x_{m+1}, x_{m+2}, \ldots, x_n$, then the uncertain process*

$$X_t = f(X_{1t}, X_{2t}, \ldots, X_{nt})$$

is a contour process with an α-path

$$X_t^\alpha = f\left(X_{1t}^\alpha, \ldots, X_{mt}^\alpha, X_{m+1,t}^{1-\alpha}, \ldots, X_{nt}^{1-\alpha}\right).$$

Proof According to the monotonicity of the function f, we have

$$\{X_t \leq f\left(X_{1t}^\alpha, \ldots, X_{mt}^\alpha, X_{m+1,t}^{1-\alpha}, \ldots, X_{nt}^{1-\alpha}\right), \forall t\}$$
$$= \{f(X_{1t}, X_{2t}, \ldots, X_{nt}) \leq f\left(X_{1t}^\alpha, \ldots, X_{mt}^\alpha, X_{m+1,t}^{1-\alpha}, \ldots, X_{nt}^{1-\alpha}\right), \forall t\}$$
$$\supset \bigcap_{i=1}^m \{X_{it} \leq X_{it}^\alpha, \forall t\} \cap \bigcap_{i=m+1}^n \{X_{it} \geq X_{it}^{1-\alpha}, \forall t\}$$

and

$$\{X_t > f\left(X_{1t}^\alpha, \ldots, X_{mt}^\alpha, X_{m+1,t}^{1-\alpha}, \ldots, X_{nt}^{1-\alpha}\right), \forall t\}$$
$$= \{f(X_{1t}, X_{2t}, \ldots, X_{nt}) > f\left(X_{1t}^\alpha, \ldots, X_{mt}^\alpha, X_{m+1,t}^{1-\alpha}, \ldots, X_{nt}^{1-\alpha}\right), \forall t\}$$

$$\supset \bigcap_{i=1}^{m} \{X_{it} > X_{it}^{\alpha}, \forall t\} \cap \bigcap_{i=m+1}^{n} \{X_{it} < X_{it}^{1-\alpha}, \forall t\}.$$

By the independence of uncertain processes and the monotonicity of uncertain measure, we have

$$\mathcal{M}\{X_t \leq X_t^{\alpha}, \forall t\}$$

$$\geq \mathcal{M}\left\{\bigcap_{i=1}^{m} \{X_{it} \leq X_{it}^{\alpha}, \forall t\} \cap \bigcap_{i=m+1}^{n} \{X_{it} \geq X_{it}^{1-\alpha}, \forall t\}\right\}$$

$$= \bigwedge_{i=1}^{m} \mathcal{M}\{X_{it} \leq X_{it}^{\alpha}, \forall t\} \wedge \bigwedge_{i=m+1}^{n} \mathcal{M}\{X_{it} \geq X_{it}^{1-\alpha}, \forall t\} = \alpha$$

and

$$\mathcal{M}\{X_t > X_t^{\alpha}, \forall t\}$$

$$\geq \mathcal{M}\left\{\bigcap_{i=1}^{m} \{X_{it} > X_{it}^{\alpha}, \forall t\} \cap \bigcap_{i=m+1}^{n} \{X_{it} < X_{it}^{1-\alpha}, \forall t\}\right\}$$

$$= \bigwedge_{i=1}^{m} \mathcal{M}\{X_{it} > X_{it}^{\alpha}, \forall t\} \wedge \bigwedge_{i=m+1}^{n} \mathcal{M}\{X_{it} < X_{it}^{1-\alpha}, \forall t\} = 1 - \alpha.$$

Since

$$\mathcal{M}\{X_t \leq X_t^{\alpha}, \forall t\} + \mathcal{M}\{X_t > X_t^{\alpha}, \forall t\} \leq 1,$$

we have

$$\mathcal{M}\{X_t \leq X_t^{\alpha}, \forall t\} = \alpha,$$

$$\mathcal{M}\{X_t > X_t^{\alpha}, \forall t\} = 1 - \alpha.$$

The theorem is proved.

Example 4.3 Let X_{1t} and X_{2t} be independent contour processes with α-paths X_{1t}^{α} and X_{2t}^{α}, respectively. Then the uncertain process $Y_t = X_{1t} + X_{2t}$ is a contour process with an α-path

$$Y_t^{\alpha} = X_{1t}^{\alpha} + X_{2t}^{\alpha},$$

and the uncertain process $Z_t = X_{1t} - X_{2t}$ is a contour process with an α-path

$$Z_t^{\alpha} = X_{1t}^{\alpha} - X_{2t}^{1-\alpha}.$$

Example 4.4 Let X_{1t} and X_{2t} be independent and positive contour processes with α-paths X_{1t}^α and X_{2t}^α, respectively. Then the uncertain process $Y_t = X_{1t} \cdot X_{2t}$ is a contour process with an α-path

$$Y_t^\alpha = X_{1t}^\alpha \cdot X_{2t}^\alpha,$$

and the uncertain process $Z_t = X_{1t}/X_{2t}$ is a contour process with an α-path

$$Z_t^\alpha = X_{1t}^\alpha / X_{2t}^{1-\alpha}.$$

A functional L on the real functions is called an increasing functional if for any two functions f and g such that $f(x) \le g(x)$ for any real number x, we always have $L(f) \le L(g)$. The supremum operator, infimum operator, and the time integral operator are all increasing functionals.

Theorem 4.8 (Yao [85]) *Let $f(x_1, x_2)$ be a strictly increasing function, and L be an increasing functional. Then for a contour process X_t with an α-path X_t^α, the uncertain process $f(L(X_t), X_t)$ is a contour process with an α-path $f\left(L(X_t^\alpha), X_t^\alpha\right)$.*

Proof According to the monotonicity of f and L, we have

$$\left\{ f(L(X_t), X_t) \le f\left(L(X_t^\alpha), X_t^\alpha\right), \forall t \right\} \supset \{X_t \le X_t^\alpha, \forall t\},$$

$$\left\{ f(L(X_t), X_t) > f\left(L(X_t^\alpha), X_t^\alpha\right), \forall t \right\} \supset \{X_t > X_t^\alpha, \forall t\}.$$

By using the monotonicity of uncertain measure, we have

$$\mathcal{M}\left\{ f(L(X_t), X_t) \le f\left(L(X_t^\alpha), X_t^\alpha\right), \forall t \right\} \ge \mathcal{M}\{X_t \le X_t^\alpha, \forall t\} = \alpha,$$

$$\mathcal{M}\left\{ f(L(X_t), X_t) > f\left(L(X_t^\alpha), X_t^\alpha\right), \forall t \right\} \ge \mathcal{M}\{X_t > X_t^\alpha, \forall t\} = 1 - \alpha.$$

Since

$$\mathcal{M}\left\{ f(L(X_t), X_t) \le f\left(L(X_t^\alpha), X_t^\alpha\right), \forall t \right\}$$

$$+ \mathcal{M}\left\{ f(L(X_t), X_t) > f\left(L(X_t^\alpha), X_t^\alpha\right), \forall t \right\} \le 1,$$

we have

$$\mathcal{M}\left\{ f(L(X_t), X_t) \le f\left(L(X_t^\alpha), X_t^\alpha\right), \forall t \right\} = \alpha,$$

$$\mathcal{M}\left\{ f(L(X_t), X_t) > f\left(L(X_t^\alpha), X_t^\alpha\right), \forall t \right\} = 1 - \alpha.$$

The theorem is proved.

Example 4.5 Let X_t be a contour process with an α-path X_t^α. Then the uncertain process

$$Y_t = \sup_{0 \le s \le t} X_s + X_t$$

is a contour process with an α-path

$$Y_t^\alpha = \sup_{0 \le s \le t} X_s^\alpha + X_t^\alpha,$$

and the uncertain process

$$Z_t = \int_0^t X_s ds + X_t$$

is a contour process with an α-path

$$Z_t^\alpha = \int_0^t X_s^\alpha ds + X_t^\alpha.$$

Chapter 5
Uncertain Calculus

Uncertain calculus deals with the integration and differentiation of uncertain processes. This chapter introduces uncertain calculus with respect to Liu process. The emphases of this chapter are on Liu integral and Liu process as well as the fundamental theorem and integration by parts.

5.1 Canonical Liu Process

Definition 5.1 (*Liu [38]*) An uncertain process C_t is called a canonical Liu process if
(i) its almost all sample paths are Lipschitz continuous,
(ii) it is a stationary independent increment process with $C_0 = 0$,
(iii) every increment $C_{s+t} - C_s$ is a normal uncertain variable with an uncertainty distribution

$$\Phi_t(x) = \left(1 + \exp\left(-\frac{\pi x}{\sqrt{3}t}\right)\right)^{-1}. \tag{5.1}$$

Theorem 5.1 (Liu [40]) *Let C_t be a canonical Liu process. Then for any time $t > 0$, the ratio*

$$\frac{C_t}{t} \sim \mathcal{N}(0, 1) \tag{5.2}$$

is a standard normal uncertain variable.

Proof Since the ratio C_t/t has an uncertainty distribution

$$\Psi(x) = \Phi_t(tx) = \left(1 + \exp\left(-\frac{\pi x}{\sqrt{3}}\right)\right)^{-1},$$

it is a normal uncertain variable with expected value 0 and variance 1. The theorem is proved.

© Springer-Verlag Berlin Heidelberg 2016
K. Yao, *Uncertain Differential Equations*,
Springer Uncertainty Research, DOI 10.1007/978-3-662-52729-0_5

Remark 5.1 Liu [40] showed that the canonical Liu process C_t satisfies $E[C_t] = 0$ and $t^2/2 \leq E[C_t^2] \leq t^2$. Iwamura and Xu [27] further showed $1.24t^4 \leq E[C_t^4] \leq 4.31t^4$. Hence ΔC_t is an infinitesimal with the same order as Δt.

Theorem 5.2 (Yao et al. [74]) *Let C_t be a canonical Liu process. Then there exists an uncertain variable K such that for each γ, $K(\gamma)$ is a Lipschitz constant of the sample path $C_t(\gamma)$, and*

$$\mathcal{M}\{K \leq x\} \geq 2\left(1 + \exp\left(-\frac{\pi x}{\sqrt{3}}\right)\right)^{-1} - 1. \tag{5.3}$$

Proof Define

$$K(\gamma) = \sup_{0 \leq t_1 < t_2} \frac{|C_{t_2}(\gamma) - C_{t_1}(\gamma)|}{t_2 - t_1}$$

for each sample path $C_t(\gamma)$. Then K is an uncertain variable on $(\Gamma, \mathcal{L}, \mathcal{M})$, and

$$|C_{t_2}(\gamma) - C_{t_1}(\gamma)| \leq K(\gamma)|t_2 - t_1|$$

for any times t_1 and t_2. That is, $K(\gamma)$ is a Lipschitz constant of the sample path $C_t(\gamma)$. Given $x \in \mathfrak{R}$, if

$$\left|\frac{dC_t}{dt}(\gamma)\right| \leq x, \quad \forall t > 0,$$

then we have

$$|C_{t_2}(\gamma) - C_{t_1}(\gamma)| \leq \int_{t_1}^{t_2} \left|\frac{dC_t}{dt}(\gamma)\right| dt \leq x(t_2 - t_1)$$

for any times t_1 and t_2, which means

$$\sup_{0 \leq t_1 < t_2} \frac{|C_{t_2}(\gamma) - C_{t_1}(\gamma)|}{t_2 - t_1} \leq x.$$

Thus

$$\sup_{0 \leq t_1 < t_2} \frac{|C_{t_2}(\gamma) - C_{t_1}(\gamma)|}{t_2 - t_1} \leq \left|\frac{dC_t}{dt}(\gamma)\right|, \quad \forall \gamma \in \Gamma$$

and

$$\mathcal{M}\{\gamma \in \Gamma \mid K(\gamma) \leq x\} = \mathcal{M}\left\{\sup_{0 \leq t_1 < t_2} \frac{|C_{t_2} - C_{t_1}|}{t_2 - t_1} \leq x\right\}$$
$$\geq \mathcal{M}\left\{\left|\frac{dC_t}{dt}\right| \leq x, \forall t > 0\right\} \geq 2\left(1 + \exp\left(-\frac{\pi x}{\sqrt{3}}\right)\right)^{-1} - 1.$$

The theorem is proved.

Theorem 5.3 (Yao et al. [74]) *Let C_t be a canonical Liu process. Then there exists an uncertain variable K such that for each γ, $K(\gamma)$ is a Lipschitz constant of the sample path $C_t(\gamma)$, and*

$$\lim_{x \to +\infty} \mathcal{M}\{K \le x\} = 1. \tag{5.4}$$

Proof Note that

$$2\left(1 + \exp\left(-\frac{\pi x}{\sqrt{3}}\right)\right)^{-1} - 1 \to 1$$

as $x \to +\infty$. The theorem follows immediately from Theorem 5.2.

5.2 Liu Integral

Definition 5.2 (*Liu [38]*) Let X_t be an uncertain process and C_t be a canonical Liu process. For any partition of closed interval $[a, b]$ with $a = t_1 < t_2 < \cdots < t_{k+1} = b$, the mesh is written as

$$\Delta = \max_{1 \le i \le k} |t_{i+1} - t_i|. \tag{5.5}$$

Then the Liu integral of X_t with respect to C_t on the interval $[a, b]$ is defined by

$$\int_a^b X_t dC_t = \lim_{\Delta \to 0} \sum_{i=1}^{k} X_{t_i} \cdot (C_{t_{i+1}} - C_{t_i}) \tag{5.6}$$

provided that the limit exists almost surely and is finite. In this case, the uncertain process X_t is said to be Liu integrable.

Remark 5.2 Assume that X_t is a Liu integrable uncertain process on an uncertainty space $(\Gamma, \mathcal{L}, \mathcal{M})$. Then for each $\gamma \in \Gamma$, the integral

$$\int_0^t X_s^*(\gamma) dC_s(\gamma)$$

is a sample path of

$$\int_0^t X_s dC_s.$$

Example 5.1 For any partition $0 = s_1 < s_2 < \cdots < s_{k+1} = t$, since

$$\int_0^t dC_s = \lim_{\Delta \to 0} \sum_{i=1}^{k} (C_{s_{i+1}} - C_{s_i}) \equiv C_t - C_0 = C_t,$$

we have

$$\int_0^t dC_s = C_t.$$

Example 5.2 For any partition $0 = s_1 < s_2 < \cdots < s_{k+1} = t$, since

$$
\begin{aligned}
C_t^2 &= \sum_{i=1}^k \left(C_{s_{i+1}}^2 - C_{s_i}^2 \right) \\
&= \sum_{i=1}^k \left(C_{s_{i+1}} - C_{s_i} \right)^2 + 2 \sum_{i=1}^k C_{s_i} \left(C_{s_{i+1}} - C_{s_i} \right) \\
&\to 0 + 2 \int_0^t C_s dC_s
\end{aligned}
$$

as $\Delta \to 0$, we have

$$\int_0^t C_s dC_t = \frac{1}{2} C_t^2.$$

Example 5.3 For any partition $0 = s_1 < s_2 < \cdots < s_{k+1} = t$, since

$$
\begin{aligned}
t C_t &= \sum_{i=1}^k \left(s_{i+1} C_{s_{i+1}} - s_i C_{s_i} \right) \\
&= \sum_{i=1}^k C_{s_{i+1}} (s_{i+1} - s_i) + \sum_{i=1}^k s_i (C_{s_{i+1}} - C_{s_i}) \\
&\to \int_0^t C_s ds + \int_0^t s dC_s
\end{aligned}
$$

as $\Delta \to 0$, we have

$$\int_0^t C_s ds + \int_0^t s dC_s = t C_t.$$

Theorem 5.4 (Liu [38]) *If X_t is a sample-continuous uncertain process on $[a, b]$, then it is Liu integrable with respect to C_t on $[a, b]$.*

Proof Since the uncertain process X_t is sample-continuous, its almost all sample paths are continuous functions of the time t. So the limit

$$\lim_{\Delta \to 0} \sum_{i=1}^k X_{t_i} (C_{t_{i+1}} - C_{t_i})$$

exists almost surely and is finite for any partition of the closed interval $[a, b]$, and the uncertain process X_t is Liu integrable on $[a, b]$. The theorem is proved.

Theorem 5.5 (Chen and Liu [3]) *Let X_t be a Liu integrable uncertain process on $[a, b]$. Then for a sample path $C_t(\gamma)$ with a Lipschitz constant $K(\gamma)$, we have*

$$\left| \int_a^b X_t(\gamma) dC_t(\gamma) \right| \leq K(\gamma) \int_a^b |X_t(\gamma)| dt. \tag{5.7}$$

Proof For a sample path $C_t(\gamma)$ with a Lipschitz constant $K(\gamma)$, we have

$$\left| \int_a^b X_t(\gamma) dC_t(\gamma) \right| \leq \int_a^b |X_t(\gamma)| \, |dC_t(\gamma)| \leq K(\gamma) \int_a^b |X_t(\gamma)| dt.$$

The theorem is proved.

Theorem 5.6 (Liu [38]) *If X_t is a Liu integrable uncertain process on $[a, b]$, then it is Liu integrable on each subinterval of $[a, b]$. Moreover, if $c \in [a, b]$, then*

$$\int_a^b X_t dC_t = \int_a^c X_t dC_t + \int_c^b X_t dC_t. \tag{5.8}$$

Proof Let $[a', b']$ be a subinterval of $[a, b]$. Since X_t is a Liu integrable uncertain process on $[a, b]$, for any partition

$$a = t_1 < \cdots < t_m = a' < t_{m+1} < \cdots < t_n = b' < t_{n+1} < \cdots < t_{k+1} = b,$$

the limit

$$\lim_{\Delta \to 0} \sum_{i=1}^k X_{t_i} (C_{t_{i+1}} - C_{t_i})$$

exists almost surely and is finite. So the limit

$$\lim_{\Delta \to 0} \sum_{i=m}^{n-1} X_{t_i} (C_{t_{i+1}} - C_{t_i})$$

exists almost surely and is finite, and the uncertain process X_t is Liu integrable on $[a', b']$. Next, for any partition of the closed interval $[a, b]$ with $t_m = c$ for some integer m, it follows from Definition 5.2 of Liu integral that

$$\int_a^b X_t dC_t = \lim_{\Delta \to 0} \sum_{i=1}^k X_{t_i} (C_{t_{i+1}} - C_{t_i})$$

$$= \lim_{\Delta \to 0} \sum_{i=1}^{m-1} X_{t_i} (C_{t_{i+1}} - C_{t_i}) + \lim_{\Delta \to 0} \sum_{i=m}^k X_{t_i} (C_{t_{i+1}} - C_{t_i})$$

$$= \int_a^c X_t dC_t + \int_c^b X_t dC_t.$$

The theorem is proved.

Theorem 5.7 (Liu [38], Linearity of Liu Integral) *Let X_t and Y_t be two Liu integrable uncertain processes on $[a, b]$. Then*

$$\int_a^b (\alpha X_t + \beta Y_t) dC_t = \alpha \int_a^b X_t dC_t + \beta \int_a^b Y_t dC_t \qquad (5.9)$$

for any real numbers α and β.

Proof For any partition of the closed interval $[a, b]$, it follows from Definition 5.2 of Liu integral that

$$\int_a^b (\alpha X_t + \beta Y_t) dC_t = \lim_{\Delta \to 0} \sum_{i=1}^k (\alpha X_{t_i} + \beta Y_{t_i})(C_{t_{i+1}} - C_{t_i})$$

$$= \lim_{\Delta \to 0} \alpha \sum_{i=1}^k X_{t_i}(C_{t_{i+1}} - C_{t_i}) + \lim_{\Delta \to 0} \beta \sum_{i=1}^k Y_{t_i}(C_{t_{i+1}} - C_{t_i})$$

$$= \alpha \int_a^b X_t dC_t + \beta \int_a^b Y_t dC_t.$$

The theorem is proved.

5.3   Liu Process

Definition 5.3 (*Chen and Ralescu [8]*) Let C_t be a canonical Liu process, μ_t be a time integrable uncertain process, and σ_t be a Liu integrable uncertain process. Then the uncertain process

$$Z_t = Z_0 + \int_0^t \mu_s ds + \int_0^t \sigma_s dC_s \qquad (5.10)$$

is called a Liu process, and it has a Liu differential

$$dZ_t = \mu_t dt + \sigma_t dC_t. \qquad (5.11)$$

Example 5.4 Since the canonical Liu process C_t satisfies

$$C_t = \int_0^t dC_s,$$

it is a Liu process and has a Liu differential dC_t.

Example 5.5 Since the uncertain process C_t^2 satisfies

$$C_t^2 = 2\int_0^t C_s dC_s,$$

it is a Liu process and has a Liu differential

$$d(C_t^2) = 2C_t dC_t.$$

Example 5.6 Since the uncertain process tC_t satisfies

$$tC_t = \int_0^t C_s ds + \int_0^t s dC_s,$$

it is a Liu process and has a Liu differential

$$d(tC_t) = C_t dt + t dC_t.$$

Theorem 5.8 (Chen and Ralescu [8]) *Liu process is a sample-continuous uncertain process.*

Proof Assume Z_t is a Liu process such that

$$Z_t = Z_0 + \int_0^t \mu_s ds + \int_0^t \sigma_s dC_s.$$

Then for each $\gamma \in \Gamma$, we have

$$|Z_{t_2}(\gamma) - Z_{t_1}(\gamma)| = \left| \int_{t_1}^{t_2} \mu_s(\gamma) ds + \int_{t_1}^{t_2} \sigma_s(\gamma) dC_s(\gamma) \right| \to 0$$

as $|t_2 - t_1| \to 0$. Hence Z_t is a sample-continuous uncertain process. The theorem is proved.

Fundamental Theorem

Theorem 5.9 (Liu [38], Fundamental Theorem) *Let C_t be a canonical Liu process, and $h(t, c)$ be a continuously differentiable function. Then the uncertain process $Z_t = h(t, C_t)$ is a Liu process, and it has a Liu differential*

$$dZ_t = \frac{\partial h}{\partial t}(t, C_t)dt + \frac{\partial h}{\partial c}(t, C_t)dC_t. \tag{5.12}$$

Proof Note that $\Delta C_t = C_{t+\Delta t} - C_t$ is an infinitesimal with the same order as Δt. By using Taylor series expansion, we get a first-order approximation

$$\Delta Z_t = \frac{\partial h}{\partial t}(t, C_t)\Delta t + \frac{\partial h}{\partial c}(t, C_t)\Delta C_t.$$

Letting $\Delta t \to 0$, we have

$$dZ_t = \frac{\partial h}{\partial t}(t, C_t)dt + \frac{\partial h}{\partial c}(t, C_t)dC_t.$$

The theorem is proved.

Example 5.7 Consider the Liu differential of the uncertain process C_t^2. In this case, we assume $h(t, c) = c^2$. Since

$$\frac{\partial h}{\partial t}(t, c) = 0, \quad \frac{\partial h}{\partial c}(t, c) = 2c,$$

we have

$$dC_t^2 = 2C_t dC_t.$$

Example 5.8 Consider the Liu differential of the uncertain process tC_t. In this case, we assume $h(t, c) = tc$. Since

$$\frac{\partial h}{\partial t}(t, c) = c, \quad \frac{\partial h}{\partial c}(t, c) = t,$$

we have

$$d(tC_t) = C_t dt + t dC_t.$$

Theorem 5.10 (Chen and Ralescu [8]) *Let X_t be a Liu process such that $dX_t = \mu_t dt + \sigma_t dC_t$, and $h(t, x)$ be a continuously differentiable function. Then the uncertain process $Z_t = h(t, X_t)$ is also a Liu process, and it has a Liu differential*

$$dZ_t = \left(\frac{\partial h}{\partial t}(t, X_t) + \mu_t \cdot \frac{\partial h}{\partial x}(t, X_t) \right) dt + \sigma_t \cdot \frac{\partial h}{\partial x}(t, X_t)dC_t. \tag{5.13}$$

Proof Note that

$$\Delta X_t = X_{t+\Delta t} - X_t = \mu_t \Delta t + \sigma_t \Delta C_t$$

is an infinitesimal with the same order as Δt. By using Taylor series expansion, we get a first-order approximation

$$\Delta Z_t = \frac{\partial h}{\partial t}(t, X_t)\Delta t + \frac{\partial h}{\partial x}(t, X_t)\Delta X_t$$
$$= \left(\frac{\partial h}{\partial t}(t, X_t) + \mu_t \cdot \frac{\partial h}{\partial x}(t, X_t)\right)\Delta t + \sigma_t \cdot \frac{\partial h}{\partial x}(t, X_t)\Delta C_t.$$

Letting $\Delta t \to 0$, we have

$$dZ_t = \left(\frac{\partial h}{\partial t}(t, X_t) + \mu_t \cdot \frac{\partial h}{\partial x}(t, X_t)\right)dt + \sigma_t \cdot \frac{\partial h}{\partial x}(t, X_t)dC_t.$$

The theorem is proved.

Theorem 5.11 (Chen and Ralescu [8]) *Let $X_{1t}, X_{2t}, \ldots, X_{nt}$ be Liu processes, and $h(t, x_1, x_2, \ldots, x_n)$ be a continuously differentiable function. Then the uncertain process $X_t = h(t, X_{1t}, X_{2t}, \ldots, X_{nt})$ has an uncertain differential*

$$dX_t = \frac{\partial h}{\partial t}(t, X_{1t}, X_{2t}, \ldots, X_{nt})dt + \sum_{i=1}^{n} \frac{\partial h}{\partial x_i}(t, X_{1t}, X_{2t}, \ldots, X_{nt})dX_{it}. \quad (5.14)$$

Proof Since the function h is continuously differentiable, by using Taylor series expansion, we get a first-order approximation

$$\Delta X_t = \frac{\partial h}{\partial t}(t, X_{1t}, X_{2t}, \ldots, X_{nt})\Delta t + \sum_{i=1}^{n} \frac{\partial h}{\partial x_i}(t, X_{1t}, X_{2t}, \ldots, X_{nt})\Delta X_{it}.$$

Letting $\Delta t \to 0$, we have

$$dX_t = \frac{\partial h}{\partial t}(t, X_{1t}, X_{2t}, \ldots, X_{nt})dt + \sum_{i=1}^{n} \frac{\partial h}{\partial x_i}(t, X_{1t}, X_{2t}, \ldots, X_{nt})dX_{it}.$$

The theorem is proved.

Integration by Parts

Theorem 5.12 (Liu [38], Integration by Parts) *Let X_t and Y_t be two Liu processes. Then*

$$d(X_t Y_t) = Y_t dX_t + X_t dY_t. \quad (5.15)$$

Proof Taking $h(t, x, y) = xy$ in Theorem 5.11, we have

$$\frac{\partial h}{\partial t}(t, x, y) = 0, \quad \frac{\partial h}{\partial x}(t, x, y) = y, \quad \frac{\partial h}{\partial y}(t, x, y) = x.$$

Then

$$d(X_t Y_t) = Y_t dX_t + X_t dY_t,$$

and the theorem is proved.

Example 5.9 Let X_{1t} and X_{2t} be two Liu processes such that $dX_{1t} = \mu_{1t}dt + \sigma_{1t}dC_t$ and $dX_{2t} = \mu_{2t}dt + \sigma_{2t}dC_t$. Then

$$
\begin{aligned}
d(X_{1t}X_{2t}) &= X_{1t}dX_{2t} + X_{2t}dX_{1t} \\
&= (\mu_{2t}X_{1t} + \mu_{1t}X_{2t})dt + (\sigma_{2t}X_{1t} + \sigma_{1t}X_{2t})dC_t.
\end{aligned}
$$

Chapter 6
Uncertain Differential Equation

Uncertain differential equation is a type of differential equations involving uncertain processes. This chapter introduces uncertain differential equations driven by Liu processes, including analytic methods, Yao–Chen formula, numerical methods, existence and uniqueness theorem, and stability theorems as well as their applications in stock markets.

6.1 Uncertain Differential Equation

Definition 6.1 (*Liu [37]*) Suppose that C_t is a canonical Liu process, and f and g are two measurable functions. Then

$$dX_t = f(t, X_t)dt + g(t, X_t)dC_t \qquad (6.1)$$

is called an uncertain differential equation. An uncertain process that satisfies (6.1) identically at each time t is called a solution of the uncertain differential equation.

Remark 6.1 The uncertain differential equation (6.1) is equivalent to the uncertain integral equation

$$X_s = X_0 + \int_0^s f(t, X_t)dt + \int_0^s g(t, X_t)dC_t.$$

Apparently, the solution of an uncertain differential equation is a Liu process.

Theorem 6.1 (Chen and Liu [3]) *Let $u_{1t}, u_{2t}, v_{1t},$ and v_{2t} be some measurable functions. Then the linear uncertain differential equation*

$$dX_t = (u_{1t}X_t + u_{2t})dt + (v_{1t}X_t + v_{2t})dC_t \qquad (6.2)$$

© Springer-Verlag Berlin Heidelberg 2016
K. Yao, *Uncertain Differential Equations*,
Springer Uncertainty Research, DOI 10.1007/978-3-662-52729-0_6

has a solution

$$X_t = U_t \cdot V_t \tag{6.3}$$

where

$$U_t = \exp\left(\int_0^t u_{1s}\,ds + \int_0^t v_{1s}\,dC_s\right), \tag{6.4}$$

$$V_t = X_0 + \int_0^t \frac{u_{2s}}{U_s}\,ds + \int_0^t \frac{v_{2s}}{U_s}\,dC_s. \tag{6.5}$$

Proof Note that

$$dU_t = u_{1t}U_t\,dt + v_{1t}U_t\,dC_t, \quad dV_t = \frac{u_{2t}}{U_t}\,dt + \frac{v_{2t}}{U_t}\,dC_t.$$

By using Theorem 5.12, we have

$$\begin{aligned}
dX_t &= U_t\,dV_t + V_t\,dU_t \\
&= U_t\left(\frac{u_{2t}}{U_t}\,dt + \frac{v_{2t}}{U_t}\,dC_t\right) + V_t(u_{1t}U_t\,dt + v_{1t}U_t\,dC_t) \\
&= (u_{2t}\,dt + v_{2t}\,dC_t) + (u_{1t}U_tV_t\,dt + v_{1t}U_tV_t\,dC_t) \\
&= (u_{2t}\,dt + v_{2t}\,dC_t) + (u_{1t}X_t\,dt + v_{1t}X_t\,dC_t) \\
&= (u_{1t}X_t + u_{2t})\,dt + (v_{1t}X_t + v_{2t})\,dC_t.
\end{aligned}$$

The theorem is proved.

Example 6.1 Consider the linear uncertain differential equation

$$dX_t = u_t\,dt + v_t\,dC_t$$

where u_t and v_t are two integrable functions. Since

$$U_t = \exp\left(\int_0^t 0\,ds + \int_0^t 0\,dC_s\right) = 1$$

and

$$V_t = X_0 + \int_0^t u_s\,ds + \int_0^t v_s\,dC_s,$$

the solution is

$$X_t = X_0 + \int_0^t u_s\,ds + \int_0^t v_s\,dC_s.$$

Example 6.2 Consider the linear uncertain differential equation

$$dX_t = u_t X_t dt + v_t X_t dC_t$$

where u_t and v_t are two integrable functions. Since

$$U_t = \exp\left(\int_0^t u_s ds + \int_0^t v_s dC_s\right)$$

and

$$V_t = X_0 + \int_0^t 0 ds + \int_0^t 0 dC_s = X_0,$$

the solution is

$$X_t = X_0 \exp\left(\int_0^t u_s ds + \int_0^t v_s dC_s\right).$$

Example 6.3 Consider the linear uncertain differential equation

$$dX_t = aX_t dt + \sigma dC_t$$

where a and σ are two real numbers. Since

$$U_t = \exp\left(\int_0^t a ds + \int_0^t 0 dC_s\right) = \exp(at)$$

and

$$V_t = X_0 + \int_0^t 0 ds + \sigma \int_0^t \exp(-as) dC_s = X_0 + \int_0^t \sigma \exp(-as) dC_s,$$

the solution is

$$X_t = \exp(at)\left(X_0 + \sigma \int_0^t \exp(-as) dC_s\right)$$

$$= X_0 \exp(at) + \sigma \int_0^t \exp(at - as) dC_s.$$

Example 6.4 Consider the linear uncertain differential equation

$$dX_t = m dt + \sigma X_t dC_t$$

where m and σ are two real numbers. Since

$$U_t = \exp\left(\int_0^t 0 ds + \int_0^t \sigma dC_s\right) = \exp(\sigma C_t)$$

and

$$V_t = X_0 + \int_0^t m \exp(-\sigma C_s)\mathrm{d}s + \int_0^t 0\mathrm{d}C_s = X_0 + m \int_0^t \exp(-\sigma C_s)\mathrm{d}s,$$

the solution is

$$X_t = \exp(\sigma C_t)\left(X_0 + m \int_0^t \exp(-\sigma C_s)\mathrm{d}s\right)$$

$$= X_0 \exp(\sigma C_t) + m \int_0^t \exp(\sigma C_t - \sigma C_s)\mathrm{d}s.$$

Example 6.5 Consider the linear uncertain differential equation

$$\mathrm{d}X_t = (aX_t + m)\mathrm{d}t + \sigma \mathrm{d}C_t \tag{6.6}$$

where m, a, and σ are some real numbers with $a \neq 0$. Since

$$U_t = \exp\left(\int_0^t a\,\mathrm{d}s + \int_0^t 0\mathrm{d}C_s\right) = \exp(at)$$

and

$$V_t = X_0 + m \int_0^t \exp(-as)\mathrm{d}s + \sigma \int_0^t \exp(-as)\mathrm{d}C_s,$$

the solution is

$$X_t = \exp(at)\left(X_0 + m \int_0^t \exp(-as)\mathrm{d}s + \sigma \int_0^t \exp(-as)\mathrm{d}C_s\right)$$

$$= -\frac{m}{a} + \exp(at)\left(X_0 + \frac{m}{a}\right) + \sigma \int_0^t \exp(at - as)\mathrm{d}C_s.$$

6.2 Analytic Methods

In this section, we introduce some analytic methods for solving some special types of (nonlinear) uncertain differential equations.

Type I

This subsection gives an analytic method to solve uncertain differential equations in the forms of

$$\mathrm{d}X_t = f(t, X_t)\mathrm{d}t + \sigma_t X_t \mathrm{d}C_t \tag{6.7}$$

or

$$dX_t = \mu_t X_t dt + g(t, X_t) dC_t. \tag{6.8}$$

Theorem 6.2 (Liu [52]) *Let $f(t, x)$ be a measurable function, and σ_t be an integrable function. Then the uncertain differential equation*

$$dX_t = f(t, X_t) dt + \sigma_t X_t dC_t \tag{6.9}$$

has a solution

$$X_t = Y_t^{-1} Z_t \tag{6.10}$$

where

$$Y_t = \exp\left(-\int_0^t \sigma_s dC_s\right) \tag{6.11}$$

and Z_t satisfies

$$dZ_t = Y_t f(t, Y_t^{-1} Z_t) dt, \quad Z_0 = X_0. \tag{6.12}$$

Proof Note that the uncertain process Y_t has a Liu differential

$$dY_t = -\exp\left(-\int_0^t \sigma_s dC_s\right) \sigma_t dC_t = -Y_t \sigma_t dC_t.$$

Then we have

$$
\begin{aligned}
d(X_t Y_t) &= X_t dY_t + Y_t dX_t \\
&= -X_t Y_t \sigma_t dC_t + Y_t f(t, X_t) dt + Y_t \sigma_t X_t dC_t \\
&= Y_t f(t, X_t) dt.
\end{aligned}
$$

Substituting $X_t Y_t$ with Z_t, and X_t with $Y_t^{-1} Z_t$, we have

$$dZ_t = Y_t f(t, Y_t^{-1} Z_t) dt.$$

Since $Y_0 = 1$, we have $Z_0 = X_0$. The theorem is proved.

Example 6.6 Consider the uncertain differential equation

$$dX_t = X_t^p dt + \sigma X_t dC_t$$

where p and σ are two real numbers with $p \neq 1$. Then we have $Y_t = \exp(-\sigma C_t)$ and

$$dZ_t = \exp(-\sigma C_t)(\exp(\sigma C_t) Z_t)^p dt = \exp((p-1)\sigma C_t) Z_t^p dt.$$

Since $p \neq 1$, we get

$$dZ_t^{1-p} = (1 - p) \exp((p - 1)\sigma C_t)dt$$

which means

$$Z_t^{1-p} = Z_0^{1-p} + (1 - p) \int_0^t \exp((p - 1)\sigma C_s)ds.$$

As a result,

$$Z_t = \left(Z_0^{1-p} + (1 - p) \int_0^t \exp((p - 1)\sigma C_s)ds \right)^{1/(1-p)}$$

$$= \left(X_0^{1-p} + (1 - p) \int_0^t \exp((p - 1)\sigma C_s)ds \right)^{1/(1-p)}.$$

By Theorem 6.2,

$$X_t = \exp(\sigma C_t) \left(X_0^{1-p} + (1 - p) \int_0^t \exp((p - 1)\sigma C_s)ds \right)^{1/(1-p)}.$$

Theorem 6.3 (Liu [52]) *Let μ_t be an integrable function, and $g(t, x)$ be a measurable function. Then the uncertain differential equation*

$$dX_t = \mu_t X_t dt + g(t, X_t)dC_t \tag{6.13}$$

has a solution

$$X_t = Y_t^{-1} Z_t \tag{6.14}$$

where

$$Y_t = \exp\left(-\int_0^t \mu_s ds \right) \tag{6.15}$$

and Z_t satisfies

$$dZ_t = Y_t g(t, Y_t^{-1} Z_t)dC_t, \quad Z_0 = X_0. \tag{6.16}$$

Proof Note that the uncertain process Y_t has a Liu differential

$$dY_t = -\exp\left(-\int_0^t \mu_s ds \right) \mu_t dt = -Y_t \mu_t dt.$$

Then we have

$$d(X_t Y_t) = X_t dY_t + Y_t dX_t$$

$$= -X_t Y_t \mu_t dt + Y_t \mu_t X_t dt + Y_t g(t, X_t)dC_t$$

$$= Y_t g(t, X_t)dC_t.$$

Substituting $X_t Y_t$ with Z_t, and X_t with $Y_t^{-1} Z_t$, we have

$$dZ_t = Y_t g(t, Y_t^{-1} Z_t) dC_t.$$

Since $Y_0 = 1$, we have $Z_0 = X_0$. The theorem is proved.

Example 6.7 Consider the uncertain differential equation

$$dX_t = \mu X_t dt + X_t^p dC_t$$

where μ and p are two real numbers with $p \neq 1$. Then we have $Y_t = \exp(-\mu t)$ and

$$dZ_t = \exp(-\mu t)(\exp(\mu t) Z_t)^p dC_t = \exp((p-1)\mu t) Z_t^p dC_t.$$

Since $p \neq 1$, we get

$$dZ_t^{1-p} = (1-p) \exp((p-1)\mu t) dC_t$$

which means

$$Z_t^{1-p} = Z_0^{1-p} + (1-p) \int_0^t \exp((p-1)\mu s) dC_s.$$

As a result,

$$
\begin{aligned}
Z_t &= \left(Z_0^{1-p} + (1-p) \int_0^t \exp((p-1)\mu s) dC_s \right)^{1/(1-p)} \\
&= \left(X_0^{1-p} + (1-p) \int_0^t \exp((p-1)\mu s) dC_s \right)^{1/(1-p)}.
\end{aligned}
$$

By Theorem 6.3,

$$X_t = \exp(\mu t) \left(X_0^{1-p} + (1-p) \int_0^t \exp((p-1)\mu s) dC_s \right)^{1/(1-p)}.$$

Type II

This subsection gives an analytic method to solve uncertain differential equations in the forms of

$$dX_t = f(t, X_t) dt + \sigma_t dC_t \tag{6.17}$$

or

$$dX_t = \mu_t dt + g(t, X_t) dC_t. \tag{6.18}$$

Theorem 6.4 (Yao [77]) *Let $f(t, x)$ be a measurable function, and σ_t be an integrable function. Then the uncertain differential equation*

$$dX_t = f(t, X_t)dt + \sigma_t dC_t \tag{6.19}$$

has a solution

$$X_t = Y_t + Z_t \tag{6.20}$$

where

$$Y_t = \int_0^t \sigma_s dC_s \tag{6.21}$$

and Z_t satisfies

$$dZ_t = f(t, Y_t + Z_t)dt, \quad Z_0 = X_0. \tag{6.22}$$

Proof Note that the uncertain process Y_t has a Liu differential

$$dY_t = \sigma_t dC_t.$$

Then we have

$$\begin{aligned} d(X_t - Y_t) &= dX_t - dY_t \\ &= f(t, X_t)dt + \sigma_t dC_t - \sigma_t dC_t \\ &= f(t, X_t)dt. \end{aligned}$$

Substituting $X_t - Y_t$ with Z_t, and X_t with $Y_t + Z_t$, we have

$$dZ_t = f(t, Y_t + Z_t)dt.$$

Since $Y_0 = 0$, we have $Z_0 = X_0$. The theorem is proved.

Example 6.8 Consider the uncertain differential equation

$$dX_t = \mu \exp(X_t)dt + \sigma dC_t$$

where μ and σ are two real numbers. In this case, we have $Y_t = \sigma C_t$ and

$$dZ_t = \mu \exp(\sigma C_t + Z_t)dt.$$

Then we get

$$d \exp(-Z_t) = -\mu \exp(\sigma C_t)dt$$

which means

$$\exp(-Z_t) = \exp(-Z_0) - \mu \int_0^t \exp(\sigma C_s)ds.$$

As a result,

$$Z_t = Z_0 - \ln\left(1 - \mu \int_0^t \exp\left(Z_0 + \sigma C_s\right) ds\right)$$
$$= X_0 - \ln\left(1 - \mu \int_0^t \exp\left(X_0 + \sigma C_s\right) ds\right).$$

By Theorem 6.4,

$$X_t = Y_t + Z_t = X_0 + \sigma C_t - \ln\left(1 - \mu \int_0^t \exp\left(X_0 + \sigma C_s\right) ds\right).$$

Theorem 6.5 (Yao [77]) *Let μ_t be an integrable function, and $g(t, x)$ be a measurable function. Then the uncertain differential equation*

$$dX_t = \mu_t dt + g(t, X_t) dC_t \tag{6.23}$$

has a solution

$$X_t = Y_t + Z_t \tag{6.24}$$

where

$$Y_t = \int_0^t \mu_s ds \tag{6.25}$$

and Z_t satisfies

$$dZ_t = g(t, Y_t + Z_t) dC_t, \quad Z_0 = X_0. \tag{6.26}$$

Proof Note that the uncertain process Y_t has a Liu differential

$$dY_t = \sigma_t dt.$$

Then we have

$$d(X_t - Y_t) = dX_t - dY_t$$
$$= \mu_t dt + g(t, X_t) dC_t - \mu_t dC_t$$
$$= g(t, X_t) dC_t.$$

Substituting $X_t - Y_t$ with Z_t, and X_t with $Y_t + Z_t$, we have

$$dZ_t = g(t, Y_t + Z_t) dC_t.$$

Since $Y_0 = 0$, we have $Z_0 = X_0$. The theorem is proved.

Example 6.9 Consider the uncertain differential equation

$$dX_t = \mu dt + \sigma \exp(X_t)dC_t$$

where μ and σ are two real numbers. In this case, we have $Y_t = \mu t$ and

$$dZ_t = \sigma \exp(\mu t + Z_t)dC_t.$$

Then we get

$$d\exp(-Z_t) = -\sigma \exp(\mu t)dC_t$$

which means

$$\exp(-Z_t) = \exp(-Z_0) - \sigma \int_0^t \exp(\mu s)dC_s.$$

As a result,

$$Z_t = Z_0 - \ln\left(1 - \sigma \int_0^t \exp(Z_0 + \mu s)\,dC_s\right)$$
$$= X_0 - \ln\left(1 - \sigma \int_0^t \exp(X_0 + \mu s)\,dC_s\right).$$

By Theorem 6.5,

$$X_t = Y_t + Z_t = X_0 + \mu t - \ln\left(1 - \sigma \int_0^t \exp(X_0 + \mu s)\,dC_s\right).$$

Remark 6.2 In addition to the above two analytic methods, Liu [50] proposed an analytic method to solve uncertain differential equations in the forms of

$$dX_t = f(t, X_t)dt + (v_{1t}X_t + v_{2t})dC_t$$

or

$$dX_t = (u_{1t}X_t + u_{2t})dt + g(t, X_t)dC_t,$$

and Wang [87] proposed an analytic method to solve uncertain differential equations in the forms of

$$dX_t = f(t, X_t)dt + \sigma_t X_t^p dC_t$$

or

$$dX_t = \mu_t X_t^p dt + g(t, X_t)dC_t.$$

6.3 Yao–Chen Formula

Theorem 6.6 (Yao–Chen Formula [75]) *The solution X_t of an uncertain differential equation*

$$dX_t = f(t, X_t)dt + g(t, X_t)dC_t \tag{6.27}$$

is a contour process with an α-path X_t^α that solves the corresponding ordinary differential equation

$$dX_t^\alpha = f(t, X_t^\alpha)dt + |g(t, X_t^\alpha)|\Phi^{-1}(\alpha)dt \tag{6.28}$$

where

$$\Phi^{-1}(\alpha) = \frac{\sqrt{3}}{\pi} \ln \frac{\alpha}{1 - \alpha}$$

is the inverse uncertainty distribution of standard normal uncertain variables. In other words,

$$\mathcal{M}\{X_t \le X_t^\alpha, \ \forall t\} = \alpha, \tag{6.29}$$

$$\mathcal{M}\{X_t > X_t^\alpha, \ \forall t\} = 1 - \alpha. \tag{6.30}$$

Proof Given $\alpha \in (0, 1)$, we divide the time interval into two parts,

$$T^+ = \left\{ t \mid g\left(t, X_t^\alpha\right) \ge 0 \right\}, \quad T^- = \left\{ t \mid g\left(t, X_t^\alpha\right) < 0 \right\}.$$

On the one hand, define

$$\Lambda_1^+ = \left\{ \gamma \left| \frac{dC_t(\gamma)}{dt} \le \Phi^{-1}(\alpha) \text{ for any } t \in T^+ \right. \right\},$$

$$\Lambda_1^- = \left\{ \gamma \left| \frac{dC_t(\gamma)}{dt} \ge \Phi^{-1}(1 - \alpha) \text{ for any } t \in T^- \right. \right\}.$$

Noting that T^+ and T^- are disjoint sets and C_t is an independent increment process, we have

$$\mathcal{M}\{\Lambda_1^+\} = \alpha, \quad \mathcal{M}\{\Lambda_1^-\} = \alpha, \quad \mathcal{M}\{\Lambda_1^+ \cap \Lambda_1^-\} = \alpha.$$

For any $\gamma \in \Lambda_1^+ \cap \Lambda_1^-$, since

$$g(t, X_t(\gamma))\frac{dC_t(\gamma)}{dt} \le |g(t, X_t^\alpha)|\Phi^{-1}(\alpha), \ \forall t,$$

we have

$$X_t(\gamma) \le X_t^\alpha, \ \forall t$$

according to the comparison theorems of ordinary differential equations. Then

$$\mathcal{M}\{X_t \leq X_t^\alpha, \forall t\} \geq \mathcal{M}\{\Lambda_1^+ \cap \Lambda_1^-\} = \alpha. \tag{6.31}$$

On the other hand, define

$$\Lambda_2^+ = \left\{\gamma \left| \frac{dC_t(\gamma)}{dt} > \Phi^{-1}(\alpha) \text{ for any } t \in T^+ \right.\right\},$$

$$\Lambda_2^- = \left\{\gamma \left| \frac{dC_t(\gamma)}{dt} < \Phi^{-1}(1-\alpha) \text{ for any } t \in T^- \right.\right\}.$$

Noting that T^+ and T^- are disjoint sets and C_t is an independent increment process, we have

$$\mathcal{M}\{\Lambda_2^+\} = 1 - \alpha, \quad \mathcal{M}\{\Lambda_2^-\} = 1 - \alpha, \quad \mathcal{M}\{\Lambda_2^+ \cap \Lambda_2^-\} = 1 - \alpha.$$

For any $\gamma \in \Lambda_2^+ \cap \Lambda_2^-$, since

$$g(t, X_t(\gamma))\frac{dC_t(\gamma)}{dt} > |g(t, X_t^\alpha)|\Phi^{-1}(\alpha), \forall t,$$

we have

$$X_t(\gamma) > X_t^\alpha, \forall t$$

according to the comparison theorems of ordinary differential equations. Then

$$\mathcal{M}\{X_t > X_t^\alpha, \forall t\} \geq \mathcal{M}\{\Lambda_2^+ \cap \Lambda_2^-\} = 1 - \alpha. \tag{6.32}$$

Since

$$\mathcal{M}\{X_t \leq X_t^\alpha, \forall t\} + \mathcal{M}\{X_t > X_t^\alpha, \forall t\} \leq 1,$$

we have

$$\mathcal{M}\{X_t \leq X_t^\alpha, \forall t\} = \alpha, \quad \mathcal{M}\{X_t > X_t^\alpha, \forall t\} = 1 - \alpha$$

from Inequalities (6.31) and (6.32). The theorem is proved.

Remark 6.3 According to Chap. 4, the inverse uncertainty distribution, expected value, extreme value, and time integral of the solution of an uncertain differential equation could all be obtained via the α-paths.

Example 6.10 The solution X_t of the uncertain differential equation

$$dX_t = udt + vdC_t, \quad X_0 = 0$$

is a contour process with an α-path

$$X_t^\alpha = ut + |v|t \cdot \frac{\sqrt{3}}{\pi} \ln \frac{\alpha}{1-\alpha}.$$

Example 6.11 The solution X_t of the uncertain differential equation

$$dX_t = uX_t dt + vX_t dC_t, \quad X_0 = 1$$

is a contour process with an α-path

$$X_t^\alpha = \exp\left(ut + |v|t \cdot \frac{\sqrt{3}}{\pi} \ln \frac{\alpha}{1-\alpha} \right).$$

Example 6.12 The solution X_t of the uncertain differential equation

$$dX_t = aX_t dt + \sigma dC_t, \quad X_0 = 0$$

is a contour process with an α-path

$$X_t^\alpha = \frac{|\sigma|}{a} \cdot \frac{\sqrt{3}}{\pi} \ln \frac{\alpha}{1-\alpha} \cdot (\exp(at) - 1).$$

Example 6.13 The solution X_t of the uncertain differential equation

$$dX_t = mdt + \sigma X_t dC_t, \quad X_0 = 0$$

is a contour process with an α-path

$$X_t^\alpha = \frac{m}{|\sigma|\Phi^{-1}(\alpha)} \cdot \left(\exp(|\sigma|\Phi^{-1}(\alpha)t) - 1 \right)$$

where

$$\Phi^{-1}(\alpha) = \frac{\sqrt{3}}{\pi} \ln \frac{\alpha}{1-\alpha}.$$

Example 6.14 The solution X_t of the uncertain differential equation

$$dX_t = (m + aX_t)dt + \sigma dC_t, \quad X_0 = 0$$

is a contour process with an α-path

$$X_t^\alpha = \left(\frac{m}{a} + \frac{|\sigma|}{a} \cdot \frac{\sqrt{3}}{\pi} \ln \frac{\alpha}{1-\alpha} \right) \cdot (\exp(at) - 1).$$

6.4 Numerical Methods

For a general uncertain differential equation, it is difficult or impossible to find its analytic solution. Even if the analytic solution is available, sometimes we cannot get its uncertainty distribution, extreme value, or time integral. Alternatively, Yao–Chen formula (Theorem 6.6) provides a numerical method to solve an uncertain differential equation via the α-paths, whose procedure is designed as follows.

Step 1 Fix α on $(0, 1)$.

Step 2 Solve the ordinary differential equation

$$\mathrm{d}X_t^\alpha = f(t, X_t^\alpha)\mathrm{d}t + |g(t, X_t^\alpha)|\Phi^{-1}(\alpha)\mathrm{d}t \tag{6.33}$$

by a numerical method where

$$\Phi^{-1}(\alpha) = \frac{\sqrt{3}}{\pi}\ln\frac{\alpha}{1-\alpha}.$$

Step 3 Obtain the α-path.

Remark 6.4 Yao and Chen [75] suggested to solve Eq. (6.33) using the Euler scheme

$$X_{(i+1)h}^\alpha = X_{ih}^\alpha + f(ih, X_{ih}^\alpha)h + |g(ih, X_{ih}^\alpha)|\Phi^{-1}(\alpha)h$$

where h is the step length. Besides, Yang and Shen [69] suggested to solve Eq. (6.33) using the Runge–Kutta scheme

$$X_{(i+1)h}^\alpha = X_{ih}^\alpha + \frac{h}{6}(k_1 + 2k_2 + 2k_3 + k_4)$$

where

$$k_1 = f(ih, X_{ih}^\alpha) + |g(ih, X_{ih}^\alpha)|\Phi^{-1}(\alpha),$$
$$k_2 = f(ih + h/2, X_{ih}^\alpha + h^2k_1/2) + |g(ih + h/2, X_{ih}^\alpha + h^2k_1/2)|\Phi^{-1}(\alpha),$$
$$k_3 = f(ih + h/2, X_{ih}^\alpha + h^2k_2/2) + |g(ih + h/2, X_{ih}^\alpha + h^2k_2/2)|\Phi^{-1}(\alpha),$$
$$k_4 = f(ih + h, X_{ih}^\alpha + h^2k_3) + |g(ih + h, X_{ih}^\alpha + h^2k_3)|\Phi^{-1}(\alpha).$$

Consider the solution of an uncertain differential equation on the interval $[0, T]$. Fix N_1 and N_2 as some large enough integers. Let α change from ϵ to $1 - \epsilon$ with the step $\epsilon = 1/N_1$. By using the above Steps 1–3 with the step $h = T/N_2$, we obtain a spectrum of α-paths in discrete form $X_{ih}^{j\epsilon}$, $i = 1, 2, \ldots, N_2$, $j = 1, 2, \ldots, N_1$. Then the uncertain variable X_T has an inverse uncertainty distribution $X_{N_2h}^{j\epsilon}$ in the discrete form and has an expected value

$$E[X_T] = \sum_{j=1}^{N_1} X_{N_2 h}^{j\epsilon}/N_1.$$

The supremum value

$$\sup_{0 \le t \le T} X_t$$

has an inverse uncertainty distribution

$$\sup_{0 \le i \le N_2} X_{ih}^{j\epsilon}$$

in the discrete form and has an expected value

$$E\left[\sup_{0 \le t \le T} X_t\right] = \sum_{j=1}^{N_1} \left(\sup_{0 \le i \le N_2} X_{ih}^{j\epsilon}\right)/N_1.$$

The infimum value

$$\inf_{0 \le t \le T} X_t$$

has an inverse uncertainty distribution

$$\inf_{0 \le i \le N_2} X_{ih}^{j\epsilon}$$

in the discrete form and has an expected value

$$E\left[\inf_{0 \le t \le T} X_t\right] = \sum_{j=1}^{N_1} \left(\inf_{0 \le i \le N_2} X_{ih}^{j\epsilon}\right)/N_1.$$

The time integral

$$\int_0^T X_t dt$$

has an inverse uncertainty distribution

$$\sum_{i=1}^{N_2} X_{ih}^{j\epsilon}/N_2$$

in the discrete form and has an expected value

$$E\left[\int_0^T X_t dt\right] = \sum_{j=1}^{N_1}\sum_{i=1}^{N_2} X_{ih}^{j\epsilon}/(N_2 N_1).$$

Example 6.15 Consider an uncertain differential equation

$$dX_t = \sin(X_t)dt + \cos(X_t)dC_t, \quad X_0 = 1.$$

We have

$$E[X_1] = 1.8678, \quad E\left[\sup_{0 \le t \le 1} X_t\right] = 1.9199,$$

$$E\left[\inf_{0 \le t \le 1} X_t\right] = 0.9479, \quad E\left[\int_0^1 X_t dt\right] = 1.4205.$$

Example 6.16 Consider an uncertain differential equation

$$dX_t = \sin(t + X_t)dt + \cos(t - X_t)dC_t, \quad X_0 = 1.$$

We have

$$E[X_1] = 1.6550, \quad E\left[\sup_{0 \le t \le 1} X_t\right] = 1.7130,$$

$$E\left[\inf_{0 \le t \le 1} X_t\right] = 0.9446, \quad E\left[\int_0^1 X_t dt\right] = 1.3853.$$

Example 6.17 Consider an uncertain differential equation

$$dX_t = (t + X_t)dt + (t^2 - X_t^2)dC_t, \quad X_0 = 0.$$

We have

$$E[X_1] = 0.6516, \quad E\left[\sup_{0 \le t \le 1} X_t\right] = 0.6657,$$

$$E\left[\inf_{0 \le t \le 1} X_t\right] = -0.0102, \quad E\left[\int_0^1 X_t dt\right] = 0.2110.$$

6.5   Existence and Uniqueness Theorem

In this section, we give a sufficient condition for an uncertain differential equation having a unique solution.

Theorem 6.7 (Chen and Liu [3]) *The uncertain differential equation*

$$dX_t = f(t, X_t)dt + g(t, X_t)dC_t \tag{6.34}$$

*with an initial value X_0 has a unique solution if the coefficients $f(t, x)$ and $g(t, x)$
satisfy the linear growth condition*

$$|f(t, x)| + |g(t, x)| \leq L(1 + |x|), \quad \forall x \in \Re, t \geq 0 \tag{6.35}$$

and the Lipschitz condition

$$|f(t, x) - f(t, y)| + |g(t, x) - g(t, y)| \leq L|x - y|, \quad \forall x, y \in \Re, t \geq 0 \tag{6.36}$$

for some constant L.

Proof We first prove that there exists a solution of (6.34) on the interval $[0, T]$ for
any given real number T by means of successive approximation. For each $\gamma \in \Gamma$,
define $X_t^{(0)}(\gamma) = X_0$,

$$X_t^{(n+1)}(\gamma) = X_0 + \int_0^t f\left(s, X_s^{(n)}(\gamma)\right) ds + \int_0^t g\left(s, X_s^{(n)}(\gamma)\right) dC_s(\gamma)$$

and

$$Q_t^{(n)}(\gamma) = \sup_{0 \leq s \leq t} \left| X_s^{(n+1)}(\gamma) - X_s^{(n)}(\gamma) \right|$$

for $n = 1, 2, \ldots$ We will prove by induction method that

$$Q_t^{(n)}(\gamma) \leq (1 + |X_0|) \frac{L^{n+1}(1 + K(\gamma))^{n+1}}{(n + 1)!} t^{n+1} \tag{6.37}$$

for almost every $\gamma \in \Gamma$ and for every nonnegative integer n, where $K(\gamma)$ is the
Lipschitz constant of $C_t(\gamma)$. Since the right term of (6.37) satisfies

$$\sum_{n=0}^{\infty} (1 + |X_0|) \frac{L^{n+1}(1 + K(\gamma))^{n+1}}{(n + 1)!} t^{n+1} < +\infty, \forall t \in [0, T],$$

it follows from the Weierstrass criterion that $X_t^{(n)}(\gamma)$ converges uniformly on $[0, T]$,
whose limit is denoted by $X_t(\gamma)$. Then we have

$$X_t(\gamma) = X_0 + \int_0^t f(s, X_s(\gamma)) ds + \int_0^t g(s, X_s(\gamma)) dC_s(\gamma).$$

Therefore, the uncertain process X_t is just the solution of the uncertain differential
equation (6.34). The inequality (6.37) is proved as follows. For $n = 0$, we have

$$Q_t^{(0)}(\gamma) = \sup_{0 \le s \le t} \left| \int_0^s f(u, X_0) du + \int_0^s g(u, X_0) dC_u(\gamma) \right|$$

$$\le \sup_{0 \le s \le t} \left| \int_0^s f(u, X_0) du \right| + \sup_{0 \le s \le t} \left| \int_0^s g(u, X_0) dC_u(\gamma) \right|$$

$$\le \sup_{0 \le s \le t} \int_0^s |f(u, X_0)| du + K(\gamma) \sup_{0 \le s \le t} \int_0^s |g(u, X_0)| du$$

$$\le \int_0^t |f(u, X_0)| du + K(\gamma) \int_0^t |g(u, X_0)| du$$

$$\le (1 + |X_0|) L (1 + K(\gamma)) t$$

by Theorems 3.5 and 5.5. Assume the inequality (6.37) holds for the integer n, i.e.,

$$Q_t^{(n)}(\gamma) = \sup_{0 \le s \le t} \left| X_s^{(n+1)}(\gamma) - X_s^{(n)}(\gamma) \right|$$

$$\le (1 + |X_0|) \frac{L^{n+1}(1 + K(\gamma))^{n+1}}{(n+1)!} t^{n+1}.$$

Then we have

$$Q_t^{(n+1)}(\gamma) = \sup_{0 \le s \le t} \left| X_s^{(n+2)}(\gamma) - X_s^{(n+1)}(\gamma) \right|$$

$$= \sup_{0 \le s \le t} \left| \int_0^s f\left(u, X_u^{(n+1)}(\gamma)\right) - f\left(u, X_u^{(n)}(\gamma)\right) du \right.$$

$$\left. + \int_0^s g\left(u, X_u^{(n+1)}(\gamma)\right) - g\left(u, X_u^{(n)}(\gamma)\right) dC_u(\gamma) \right|$$

$$\le \sup_{0 \le s \le t} \left| \int_0^s f\left(u, X_u^{(n+1)}(\gamma)\right) - f\left(u, X_u^{(n)}(\gamma)\right) du \right|$$

$$+ \sup_{0 \le s \le t} \left| \int_0^s g\left(u, X_u^{(n+1)}(\gamma)\right) - g\left(u, X_u^{(n)}(\gamma)\right) dC_u(\gamma) \right|$$

$$\le \int_0^t \left| f\left(u, X_u^{(n+1)}(\gamma)\right) - f\left(u, X_u^{(n)}(\gamma)\right) \right| du$$

$$+ K(\gamma) \int_0^t \left| g\left(u, X_u^{(n+1)}(\gamma)\right) - g\left(u, X_u^{(n)}(\gamma)\right) \right| du$$

$$\le L \int_0^t \left| X_u^{(n+1)}(\gamma) - X_u^{(n)}(\gamma) \right| du + K(\gamma) L \int_0^t \left| X_u^{(n+1)}(\gamma) - X_u^{(n)}(\gamma) \right| du$$

$$\le L(1 + K(\gamma)) \int_0^t (1 + |X_0|) \frac{L^{n+1}(1 + K(\gamma))^{n+1}}{(n+1)!} u^{n+1} du$$

$$= (1 + |X_0|) \frac{L^{n+2}(1 + K(\gamma))^{n+2}}{(n+2)!} t^{n+2}.$$

It means the inequality (6.37) also holds for the integer $n + 1$. Thus the inequality (6.37) holds for all nonnegative integers.

Next, we prove the uniqueness of the solution under the given conditions. Assume that X_t and X_t^* are two solutions of the uncertain differential equation (6.34) with a common initial value X_0. Then for almost every $\gamma \in \Gamma$, we have

$$
\begin{aligned}
&|X_t(\gamma) - X_t^*(\gamma)| \\
&= \left| \int_0^t f(s, X_s(\gamma)) - f(s, X_s^*(\gamma))ds + \int_0^t g(s, X_s(\gamma)) - g(s, X_s^*(\gamma))dC_s(\gamma) \right| \\
&\leq \left| \int_0^t f(s, X_s(\gamma)) - f(s, X_s^*(\gamma))ds \right| + \left| \int_0^t g(s, X_s(\gamma)) - g(s, X_s^*(\gamma))dC_s(\gamma) \right| \\
&\leq \int_0^t \left| f(s, X_s(\gamma)) - f(s, X_s^*(\gamma)) \right| ds + K(\gamma) \int_0^t \left| g(s, X_s(\gamma)) - g(s, X_s^*(\gamma)) \right| ds \\
&\leq L \int_0^t \left| X_s(\gamma) - X_s^*(\gamma) \right| ds + K(\gamma)L \int_0^t \left| X_s(\gamma) - X_s^*(\gamma) \right| ds \\
&= L(1 + K(\gamma)) \int_0^t \left| X_s(\gamma) - X_s^*(\gamma) \right| ds.
\end{aligned}
$$

By the Grönwall's inequality, we obtain

$$
|X_t(\gamma) - X_t^*(\gamma)| \leq 0 \cdot \exp(L(1 + K(\gamma))t) = 0.
$$

That means $X_t = X_t^*$ almost surely. The uniqueness of the solution is verified. The theorem is proved.

6.6 Stability Theorems

In this section, we introduce the concepts of stability in three senses, namely stability in measure, stability in mean, and almost sure stability.

Stability in Measure

Definition 6.2 *(Liu [38])* An uncertain differential equation

$$
dX_t = f(t, X_t)dt + g(t, X_t)dC_t \tag{6.38}
$$

is said to be stable in measure if for any two solutions X_t and Y_t with different initial values X_0 and Y_0, we have

$$
\lim_{|X_0 - Y_0| \to 0} \mathcal{M} \left\{ \sup_{t \geq 0} |X_t - Y_t| \leq \varepsilon \right\} = 1 \tag{6.39}
$$

for any given number $\varepsilon > 0$.

Example 6.18 Consider the uncertain differential equation

$$dX_t = \mu dt + \sigma dC_t. \tag{6.40}$$

Since its solutions with different initial values X_0 and Y_0 are

$$X_t = X_0 + \mu t + \sigma C_t,$$

$$Y_t = Y_0 + \mu t + \sigma C_t,$$

respectively, we have

$$\sup_{t \geq 0} |X_t - Y_t| = |X_0 - Y_0|$$

almost surely. Then

$$\lim_{|X_0 - Y_0| \to 0} \mathcal{M} \left\{ \sup_{t \geq 0} |X_t - Y_t| \leq \varepsilon \right\} = \lim_{|X_0 - Y_0| \to 0} \mathcal{M} \{ |X_0 - Y_0| \leq \varepsilon \} = 1,$$

and the uncertain differential equation (6.40) is stable in measure.

Example 6.19 Consider the uncertain differential equation

$$dX_t = X_t dt + \sigma dC_t. \tag{6.41}$$

Since its solutions with different initial values X_0 and Y_0 are

$$X_t = \exp(t) X_0 + \sigma \exp(t) \int_0^t \exp(-s) dC_s,$$

$$Y_t = \exp(t) Y_0 + \sigma \exp(t) \int_0^t \exp(-s) dC_s,$$

respectively, we have

$$\sup_{t \geq 0} |X_t - Y_t| = \sup_{t \geq 0} |X_0 - Y_0| \cdot \exp(t) = +\infty$$

almost surely. Then

$$\lim_{|X_0 - Y_0| \to 0} \mathcal{M} \left\{ \sup_{t \geq 0} |X_t - Y_t| \leq \varepsilon \right\}$$

$$= \lim_{|X_0 - Y_0| \to 0} \mathcal{M} \left\{ \sup_{t \geq 0} |X_0 - Y_0| \cdot \exp(t) \leq \varepsilon \right\} = 0,$$

and the uncertain differential equation (6.41) is not stable in measure.

Theorem 6.8 (Yao et al. [74]) *Assume the uncertain differential equation*

$$\mathrm{d}X_t = f(t, X_t)\mathrm{d}t + g(t, X_t)\mathrm{d}C_t \qquad (6.42)$$

has a unique solution for each given initial value. Then it is stable in measure if the coefficients $f(t, x)$ and $g(t, x)$ satisfy the strong Lipschitz condition

$$|f(t, x) - f(t, y)| + |g(t, x) - g(t, y)| \le L_t|x - y|, \quad \forall x, y \in \Re, t \ge 0 \quad (6.43)$$

where L_t is some positive function satisfying

$$\int_0^{+\infty} L_t\mathrm{d}t < +\infty. \qquad (6.44)$$

Proof Let X_t and Y_t be the solutions of the uncertain differential equation (6.42) with different initial values X_0 and Y_0, respectively. Then for a Lipschitz continuous sample path $C_t(\gamma)$, we have

$$X_t(\gamma) = X_0 + \int_0^t f(s, X_s(\gamma))\mathrm{d}s + \int_0^t g(s, X_s(\gamma))\mathrm{d}C_s(\gamma),$$

$$Y_t(\gamma) = Y_0 + \int_0^t f(s, Y_s(\gamma))\mathrm{d}s + \int_0^t g(s, Y_s(\gamma))\mathrm{d}C_s(\gamma).$$

By the strong Lipschitz condition, we have

$$|X_t(\gamma) - Y_t(\gamma)| \le |X_0 - Y_0| + \int_0^t |f(s, X_s(\gamma)) - f(s, Y_s(\gamma))|\mathrm{d}s$$

$$+ \int_0^t |g(s, X_s(\gamma)) - g(s, Y_s(\gamma))||\mathrm{d}C_s(\gamma)|$$

$$\le |X_0 - Y_0| + \int_0^t L_s|X_s(\gamma) - Y_s(\gamma)|\mathrm{d}s$$

$$+ \int_0^t K(\gamma)L_s|X_s(\gamma) - Y_s(\gamma)|\mathrm{d}s$$

$$= |X_0 - Y_0| + (1 + K(\gamma)) \int_0^t L_s|X_s(\gamma) - Y_s(\gamma)|\mathrm{d}s$$

where $K(\gamma)$ is the Lipschitz constant of $C_t(\gamma)$. It follows from the Grönwall's inequality that

$$|X_t(\gamma) - Y_t(\gamma)| \le |X_0 - Y_0| \cdot \exp\left((1 + K(\gamma)) \int_0^t L_s\mathrm{d}s\right)$$

$$\le |X_0 - Y_0| \cdot \exp\left((1 + K(\gamma)) \int_0^{+\infty} L_s\mathrm{d}s\right)$$

for any $t \geq 0$. Thus we have

$$\sup_{t \geq 0} |X_t - Y_t| \leq |X_0 - Y_0| \cdot \exp \left((1 + K) \int_0^{+\infty} L_s \mathrm{d}s \right)$$

almost surely, where K is a nonnegative uncertain variable such that

$$\lim_{x \to \infty} \mathcal{M}\{\gamma \in \Gamma \mid K(\gamma) \leq x\} = 1$$

by Theorem 5.3. For any given $\epsilon > 0$, there exists a real number H such that $\mathcal{M}\{\gamma \mid K(\gamma) \leq H\} \geq 1 - \epsilon$. Take

$$\delta = \exp \left(-(1 + H) \int_0^{+\infty} L_s \mathrm{d}s \right) \epsilon.$$

Then $|X_t(\gamma) - Y_t(\gamma)| \leq \epsilon$ for any time t provided that $|X_0 - Y_0| \leq \delta$ and $K(\gamma) \leq H$. It means

$$\mathcal{M} \left\{ \sup_{t \geq 0} |X_t - Y_t| \leq \epsilon \right\} > 1 - \epsilon$$

as long as $|X_0 - Y_0| \leq \delta$. In other words,

$$\lim_{|X_0 - Y_0| \to 0} \mathcal{M} \left\{ \sup_{t \geq 0} |X_t - Y_t| \leq \epsilon \right\} = 1,$$

and the uncertain differential equation (6.42) is stable in measure. The theorem is proved.

Example 6.20 Consider a nonlinear uncertain differential equation

$$\mathrm{d}X_t = \exp \left(-t^2 \right) X_t \mathrm{d}t + \exp \left(-t - X_t^2 \right) \mathrm{d}C_t. \tag{6.45}$$

Note that $f(t, x) = \exp \left(-t^2 \right) x$ and $g(t, x) = \exp \left(-t - x^2 \right)$ satisfy the strong Lipschitz condition

$$|f(t, x) - f(t, y)| \leq \exp \left(-t^2 \right) |x - y|, \quad \forall x, y \in \Re, t \geq 0$$

$$|g(t, x) - g(t, y)| \leq \exp(-t)|x - y|, \quad \forall x, y \in \Re, t \geq 0$$

with

$$\int_0^{+\infty} \exp \left(-t^2 \right) + \exp \left(-t \right) \mathrm{d}t < +\infty.$$

By using Theorem 6.8, the nonlinear uncertain differential equation (6.45) is stable in measure.

In fact, Theorem 6.8 gives a sufficient but not necessary condition for an uncertain differential equation being stable in measure.

Example 6.21 Consider the uncertain differential equation

$$dX_t = -X_t dt + \sigma dC_t. \tag{6.46}$$

Since its solutions with different initial values X_0 and Y_0 are

$$X_t = \exp(-t)X_0 + \sigma \exp(-t) \int_0^t \exp(s) dC_s,$$

$$Y_t = \exp(-t)Y_0 + \sigma \exp(-t) \int_0^t \exp(s) dC_s,$$

respectively, we have

$$\sup_{t \geq 0} |X_t - Y_t| = \sup_{t \geq 0} |X_0 - Y_0| \cdot \exp(-t) = |X_0 - Y_0|$$

almost surely. Then

$$\lim_{|X_0 - Y_0| \to 0} \mathcal{M} \left\{ \sup_{t \geq 0} |X_t - Y_t| \leq \varepsilon \right\} = \lim_{|X_0 - Y_0| \to 0} \mathcal{M} \{|X_0 - Y_0| \leq \varepsilon\} = 1,$$

and the uncertain differential equation (6.46) is stable in measure. However, the coefficient $f(t, x) = -x$ does not satisfy the strong Lipschitz condition in Theorem 6.8.

Stability in Mean

Definition 6.3 (*Yao et al. [82]*) An uncertain differential equation

$$dX_t = f(t, X_t)dt + g(t, X_t)dC_t \tag{6.47}$$

is said to be stable in mean if for any two solutions X_t and Y_t with different initial values X_0 and Y_0, we have

$$\lim_{|X_0 - Y_0| \to 0} E \left[\sup_{t \geq 0} |X_t - Y_t| \right] = 0. \tag{6.48}$$

Example 6.22 Consider the uncertain differential equation

$$dX_t = \mu dt + \sigma dC_t. \tag{6.49}$$

Since its solutions with different initial values X_0 and Y_0 are

$$X_t = X_0 + \mu t + \sigma C_t,$$

$$Y_t = Y_0 + \mu t + \sigma C_t,$$

respectively, we have

$$\sup_{t \geq 0} |X_t - Y_t| = |X_0 - Y_0|$$

almost surely. Then

$$\lim_{|X_0 - Y_0| \to 0} E\left[\sup_{t \geq 0} |X_t - Y_t|\right] = \lim_{|X_0 - Y_0| \to 0} E[|X_0 - Y_0|] = 0,$$

and the uncertain differential equation (6.49) is stable in mean.

Example 6.23 Consider the uncertain differential equation

$$dX_t = X_t dt + \sigma dC_t. \tag{6.50}$$

Since its solutions with different initial values X_0 and Y_0 are

$$X_t = \exp(t)X_0 + \sigma \exp(t) \int_0^t \exp(-s)dC_s,$$

$$Y_t = \exp(t)Y_0 + \sigma \exp(t) \int_0^t \exp(-s)dC_s,$$

respectively, we have

$$\sup_{t \geq 0} |X_t - Y_t| = \sup_{t \geq 0} |X_0 - Y_0| \cdot \exp(t) = +\infty$$

almost surely. Then

$$\lim_{|X_0 - Y_0| \to 0} E\left[\sup_{t \geq 0} |X_t - Y_t|\right]$$

$$= \lim_{|X_0 - Y_0| \to 0} E\left[\sup_{t \geq 0} |X_0 - Y_0| \cdot \exp(t)\right] = +\infty,$$

and the uncertain differential equation (6.50) is not stable in mean.

Theorem 6.9 (Yao et al. [82]) *If an uncertain differential equation is stable in mean, then it is stable in measure.*

Proof Let X_t and Y_t be the solutions of an uncertain differential equation with different initial values X_0 and Y_0, respectively. Then it follows from Definition 6.3 of stability in mean that

$$\lim_{|X_0 - Y_0| \to 0} E\left[\sup_{t \geq 0} |X_t - Y_t|\right] = 0.$$

By using Markov inequality (Theorem 2.10), for any given real number $\varepsilon > 0$, we have

$$\lim_{|X_0 - Y_0| \to 0} \mathcal{M}\left\{\sup_{t \geq 0} |X_t - Y_t| \geq \varepsilon\right\} \leq \lim_{|X_0 - Y_0| \to 0} E\left[\sup_{t \geq 0} |X_t - Y_t|\right]\bigg/\varepsilon = 0.$$

Thus it follows from Definition 6.2 of stability in measure that the uncertain differential equation is stable in measure. The theorem is proved.

Remark 6.5 Generally, stability in measure does not imply stability in mean. Consider the uncertain differential equation

$$dX_t = \frac{2X_t}{(t+1)^2} dC_t. \tag{6.51}$$

Obviously, the coefficients $f(t, x) = 0$ and $g(t, x) = 2x/(t+1)^2$ satisfy the strong Lipschitz condition in Theorem 6.8, so the uncertain differential equation (6.51) is stable in measure. It is easy to verify that the uncertain differential equation (6.51) has a solution

$$X_t \equiv 0$$

with an initial value $X_0 = 0$ and has a solution

$$Y_t = Y_0 \exp\left(\int_0^t \frac{2}{(s+1)^2} dC_s\right)$$

with an initial value $Y_0 \neq 0$. Then

$$\sup_{t \geq 0} |X_t - Y_t| = |Y_0| \cdot \sup_{t \geq 0} \exp\left(\int_0^t \frac{2}{(s+1)^2} dC_s\right)$$

almost surely, and

$$E\left[\sup_{t \geq 0} |X_t - Y_t|\right] = |Y_0| \cdot E\left[\sup_{t \geq 0} \exp\left(\int_0^t \frac{2}{(s+1)^2} dC_s\right)\right]$$

$$\geq |Y_0| \cdot E\left[\exp\left(\int_0^{+\infty} \frac{2}{(s+1)^2} dC_s\right)\right].$$

Since

$$\int_0^{+\infty} \frac{2}{(s+1)^2} \mathrm{d}C_s \sim \mathcal{N}\left(0, \int_0^{+\infty} \frac{2}{(s+1)^2} \mathrm{d}s\right) = \mathcal{N}(0, 2),$$

we have

$$E\left[\exp\left(\int_0^{+\infty} \frac{2}{(s+1)^2} \mathrm{d}C_s\right)\right] = \infty.$$

As a result,

$$E\left[\sup_{t \geq 0} |X_t - Y_t|\right] = \infty$$

provided that $Y_0 \neq 0$. Thus the uncertain differential equation (6.51) is not stable in mean.

Theorem 6.10 (Yao et al. [82]) *Assume the uncertain differential equation*

$$\mathrm{d}X_t = f(t, X_t)\mathrm{d}t + g(t, X_t)\mathrm{d}C_t \tag{6.52}$$

has a unique solution for each given initial value. Then it is stable in mean if the coefficients $f(t, x)$ and $g(t, x)$ satisfy the strong Lipschitz condition

$$|f(t, x) - f(t, y)| \leq L_{1t}|x - y|, \quad \forall x, y \in \mathfrak{R}, t \geq 0 \tag{6.53}$$

$$|g(t, x) - g(t, y)| \leq L_{2t}|x - y|, \quad \forall x, y \in \mathfrak{R}, t \geq 0 \tag{6.54}$$

where L_{1t} and L_{2t} are some positive functions satisfying

$$\int_0^{+\infty} L_{1t}\mathrm{d}t < +\infty, \quad \int_0^{+\infty} L_{2t}\mathrm{d}t < \frac{\pi}{\sqrt{3}}. \tag{6.55}$$

Proof Let X_t and Y_t be the solutions of the uncertain differential equation (6.52) with different initial values X_0 and Y_0, respectively. Then for a Lipschitz continuous sample path $C_t(\gamma)$, we have

$$X_t(\gamma) = X_0 + \int_0^t f(s, X_s(\gamma))\mathrm{d}s + \int_0^t g(s, X_s(\gamma))\mathrm{d}C_s(\gamma),$$

$$Y_t(\gamma) = Y_0 + \int_0^t f(s, Y_s(\gamma))\mathrm{d}s + \int_0^t g(s, Y_s(\gamma))\mathrm{d}C_s(\gamma).$$

By the strong Lipschitz condition, we have

$$|X_t(\gamma) - Y_t(\gamma)| \leq |X_0 - Y_0| + \int_0^t |f(s, X_s(\gamma)) - f(s, Y_s(\gamma))|\mathrm{d}s$$

$$+ \int_0^t |g(s, X_s(\gamma)) - g(s, Y_s(\gamma))||dC_s(\gamma)|$$

$$\leq |X_0 - Y_0| + \int_0^t L_{1s}|X_s(\gamma) - Y_s(\gamma)|ds$$

$$+ \int_0^t K(\gamma)L_{2s}|X_s(\gamma) - Y_s(\gamma)|ds$$

$$= |X_0 - Y_0| + \int_0^t (L_{1s} + K(\gamma)L_{2s})|X_s(\gamma) - Y_s(\gamma)|ds$$

where $K(\gamma)$ is the Lipschitz constant of $C_t(\gamma)$. It follows from the Grönwall's inequality that

$$|X_t(\gamma) - Y_t(\gamma)| \leq |X_0 - Y_0| \cdot \exp\left(\int_0^t L_{1s}ds\right) \cdot \exp\left(K(\gamma)\int_0^t L_{2s}ds\right)$$

$$\leq |X_0 - Y_0| \cdot \exp\left(\int_0^{+\infty} L_{1s}ds\right) \cdot \exp\left(K(\gamma)\int_0^{+\infty} L_{2s}ds\right)$$

for any $t \geq 0$. Thus we have

$$\sup_{t \geq 0} |X_t - Y_t| \leq |X_0 - Y_0| \cdot \exp\left(\int_0^{+\infty} L_{1s}ds\right) \cdot \exp\left(K\int_0^{+\infty} L_{2s}ds\right)$$

almost surely, where K is a nonnegative uncertain variable such that

$$\mathcal{M}\{\gamma \in \Gamma \mid K(\gamma) \geq x\} = 1 - \mathcal{M}\{\gamma \in \Gamma \mid K(\gamma) < x\}$$

$$\leq 2\left(1 + \exp\left(\frac{\pi x}{\sqrt{3}}\right)\right)^{-1}$$

by Theorem 5.2. Taking expected value on both sides, we have

$$E\left[\sup_{t \geq 0} |X_t(\gamma) - Y_t(\gamma)|\right]$$

$$\leq |X_0 - Y_0| \cdot \exp\left(\int_0^{+\infty} L_{1s}ds\right) \cdot E\left[\exp\left(K\int_0^{+\infty} L_{2s}ds\right)\right].$$

Since

$$\int_0^{+\infty} L_{1s}ds < +\infty,$$

we immediately have

$$\exp\left(\int_0^{+\infty} L_{1s}ds\right) < +\infty.$$

As for

$$E\left[\exp\left(K\int_0^{+\infty} L_{2s}\,ds\right)\right],$$

we write

$$r = \int_0^{+\infty} L_{2s}\,ds < \frac{\pi}{\sqrt{3}}$$

for simplicity. It follows from Definition 2.9 of expected value that

$$E[\exp(rK)] = \int_0^{+\infty} \mathcal{M}\{\exp(rK) \geq x\}dx = \int_0^{+\infty} \mathcal{M}\left\{K \geq \frac{\ln x}{r}\right\}dx$$

$$\leq 2\int_0^{+\infty} \frac{1}{1+x^{\pi/(\sqrt{3}r)}}dx < +\infty.$$

So

$$\lim_{|X_0 - Y_0| \to 0} E\left[\sup_{t \geq 0} |X_t - Y_t|\right] = 0,$$

and the uncertain differential equation (6.52) is stable in mean. The theorem is proved.

Example 6.24 Consider a nonlinear uncertain differential equation

$$dX_t = \exp\left(-t^2\right)X_t dt + \exp\left(-t - X_t^2\right)dC_t. \tag{6.56}$$

Note that $f(t, x) = \exp\left(-t^2\right)x$ and $g(t, x) = \exp\left(-t - x^2\right)$ satisfy the strong Lipschitz condition

$$|f(t, x) - f(t, y)| \leq \exp\left(-t^2\right)|x - y|, \quad \forall x, y \in \Re, t \geq 0$$

$$|g(t, x) - g(t, y)| \leq \exp(-t)|x - y|, \quad \forall x, y \in \Re, t \geq 0$$

with

$$\int_0^{+\infty} \exp\left(-t^2\right)dt < +\infty, \quad \int_0^{+\infty} \exp\left(-t\right)dt = 1 < \frac{\pi}{\sqrt{3}}.$$

By using Theorem 6.10, the nonlinear uncertain differential equation (6.56) is stable in mean.

In fact, Theorem 6.10 gives a sufficient but not necessary condition for an uncertain differential equation being stable in mean.

Example 6.25 Consider the uncertain differential equation

$$dX_t = -X_t dt + \sigma dC_t. \tag{6.57}$$

Since its solutions with different initial values X_0 and Y_0 are

$$X_t = \exp(-t)X_0 + \sigma \exp(-t) \int_0^t \exp(s) dC_s,$$

$$Y_t = \exp(-t)Y_0 + \sigma \exp(-t) \int_0^t \exp(s) dC_s,$$

respectively, we have

$$\sup_{t \geq 0} |X_t - Y_t| = \sup_{t \geq 0} |X_0 - Y_0| \cdot \exp(-t) = |X_0 - Y_0|$$

almost surely. Then

$$\lim_{|X_0 - Y_0| \to 0} E\left[\sup_{t \geq 0} |X_t - Y_t|\right] = \lim_{|X_0 - Y_0| \to 0} E[|X_0 - Y_0|] = 0,$$

and the uncertain differential equation (6.57) is stable in mean. However, the coefficient $f(t, x) = -x$ does not satisfy the strong Lipschitz condition in Theorem 6.10.

Almost Sure Stability

Definition 6.4 (*Liu et al. [49]*) An uncertain differential equation

$$dX_t = f(t, X_t)dt + g(t, X_t)dC_t \tag{6.58}$$

is said to be almost surely stable if for any two solutions X_t and Y_t with different initial values X_0 and Y_0, we have

$$\mathcal{M}\left\{\lim_{|X_0 - Y_0| \to 0} \sup_{t \geq 0} |X_t - Y_t| = 0\right\} = 1. \tag{6.59}$$

Example 6.26 Consider the uncertain differential equation

$$dX_t = \mu dt + \sigma dC_t. \tag{6.60}$$

Since its solutions with different initial values X_0 and Y_0 are

$$X_t = X_0 + \mu t + \sigma C_t,$$

$$Y_t = Y_0 + \mu t + \sigma C_t,$$

respectively, we have

$$\sup_{t \geq 0} |X_t(\gamma) - Y_t(\gamma)| = |X_0 - Y_0|, \quad \forall \gamma \in \Gamma.$$

Then

$$\mathcal{M}\left\{\lim_{|X_0-Y_0|\to 0}\sup_{t\ge 0}|X_t - Y_t| = 0\right\} = \mathcal{M}\left\{\lim_{|X_0-Y_0|\to 0}|X_0 - Y_0| = 0\right\} = 1,$$

and the uncertain differential equation (6.60) is almost surely stable.

Example 6.27 Consider the uncertain differential equation

$$\mathrm{d}X_t = X_t\mathrm{d}t + \sigma\mathrm{d}C_t. \tag{6.61}$$

Since its solutions with different initial values X_0 and Y_0 are

$$X_t = \exp(t)X_0 + \sigma\exp(t)\int_0^t \exp(-s)\mathrm{d}C_s,$$

$$Y_t = \exp(t)Y_0 + \sigma\exp(t)\int_0^t \exp(-s)\mathrm{d}C_s,$$

respectively, we have

$$\sup_{t\ge 0}|X_t(\gamma) - Y_t(\gamma)| = \sup_{t\ge 0}|X_0 - Y_0| \cdot \exp(t) = \infty, \quad \forall\gamma \in \Gamma.$$

Then

$$\mathcal{M}\left\{\lim_{|X_0-Y_0|\to 0}\sup_{t\ge 0}|X_t - Y_t| = 0\right\} = 0,$$

and the uncertain differential equation (6.61) is not almost surely stable.

Theorem 6.11 (Liu et al. [49]) *Assume the uncertain differential equation*

$$\mathrm{d}X_t = f(t, X_t)\mathrm{d}t + g(t, X_t)\mathrm{d}C_t \tag{6.62}$$

has a unique solution for each given initial value. Then it is almost surely stable if the coefficients $f(t, x)$ and $g(t, x)$ satisfy the strong Lipschitz condition

$$|f(t, x) - f(t, y)| + |g(t, x) - g(t, y)| \le L_t|x - y|, \quad \forall x, y \in \Re, t \ge 0 \tag{6.63}$$

where L_t is some positive function satisfying

$$\int_0^{+\infty} L_t\mathrm{d}t < +\infty. \tag{6.64}$$

Proof Let X_t and Y_t be the solutions of the uncertain differential equation with different initial values X_0 and Y_0, respectively. Then for a Lipschitz continuous sample path $C_t(\gamma)$, we have

$$X_t(\gamma) = X_0 + \int_0^t f(s, X_s(\gamma))\mathrm{d}s + \int_0^t g(s, X_s(\gamma))\mathrm{d}C_s(\gamma),$$

$$Y_t(\gamma) = Y_0 + \int_0^t f(s, Y_s(\gamma))\mathrm{d}s + \int_0^t g(s, Y_s(\gamma))\mathrm{d}C_s(\gamma).$$

By the strong Lipschitz condition, we have

$$|X_t(\gamma) - Y_t(\gamma)| \le |X_0 - Y_0| + \int_0^t |f(s, X_s(\gamma)) - f(s, Y_s(\gamma))|\mathrm{d}s$$

$$+ \int_0^t |g(s, X_s(\gamma)) - g(s, Y_s(\gamma))||\mathrm{d}C_s(\gamma)|$$

$$\le |X_0 - Y_0| + L_s \int_0^t |X_s(\gamma) - Y_s(\gamma)|\mathrm{d}s$$

$$+ K(\gamma)L_s \int_0^t |X_s(\gamma) - Y_s(\gamma)|\mathrm{d}s$$

$$= |X_0 - Y_0| + (1 + K(\gamma)) \int_0^t L_s|X_s(\gamma) - Y_s(\gamma)|\mathrm{d}s$$

where $K(\gamma)$ is the Lipschitz constant of $C_t(\gamma)$. It follows from the Grönwall's inequality that

$$|X_t(\gamma) - Y_t(\gamma)| \le |X_0 - Y_0| \cdot \exp\left((1 + K(\gamma)) \int_0^t L_s\mathrm{d}s\right)$$

$$\le |X_0 - Y_0| \cdot \exp\left((1 + K(\gamma)) \int_0^{+\infty} L_s\mathrm{d}s\right)$$

for any $t \ge 0$. Thus we have

$$\sup_{t \ge 0} |X_t(\gamma) - Y_t(\gamma)| \le |X_0 - Y_0| \cdot \exp\left((1 + K(\gamma)) \int_0^{+\infty} L_s\mathrm{d}s\right), \quad \forall \gamma \in \Gamma.$$

Since C_t is a Lipschitz continuous uncertain process, the uncertain variable K is almost surely finite. Then we have

$$\mathcal{M}\left\{\lim_{|X_0 - Y_0| \to 0} \sup_{t \ge 0} |X_t - Y_t| = 0\right\}$$

$$\ge \mathcal{M}\left\{\lim_{|X_0 - Y_0| \to 0} |X_0 - Y_0| \cdot \exp\left((1 + K) \int_0^{+\infty} L_s\mathrm{d}s\right) = 0\right\} = 1,$$

and the uncertain differential equation (6.62) is almost surely stable. The theorem is proved.

Example 6.28 Consider the nonlinear uncertain differential equation

$$dX_t = \exp\left(-t^2\right) X_t dt + \exp\left(-t - X_t^2\right) dC_t. \tag{6.65}$$

Note that $f(t, x) = \exp\left(-t^2\right) x$ and $g(t, x) = \exp\left(-t - x^2\right)$ satisfy the strong Lipschitz condition

$$|f(t, x) - f(t, y)| \le \exp\left(-t^2\right) |x - y|, \quad \forall x, y \in \Re, t \ge 0$$

$$|g(t, x) - g(t, y)| \le \exp(-t)|x - y|, \quad \forall x, y \in \Re, t \ge 0$$

with

$$\int_0^{+\infty} \exp\left(-t^2\right) + \exp\left(-t\right) dt < +\infty.$$

By using Theorem 6.11, the nonlinear uncertain differential equation (6.65) is almost surely stable.

In fact, Theorem 6.11 gives a sufficient but not necessary condition for an uncertain differential equation being almost surely stable.

Example 6.29 Consider the uncertain differential equation

$$dX_t = -X_t dt + \sigma dC_t. \tag{6.66}$$

Since its solutions with different initial values X_0 and Y_0 are

$$X_t = \exp(-t)X_0 + \sigma \exp(-t) \int_0^t \exp(s)dC_s,$$

$$Y_t = \exp(-t)Y_0 + \sigma \exp(-t) \int_0^t \exp(s)dC_s,$$

respectively, we have

$$\sup_{t \ge 0} |X_t(\gamma) - Y_t(\gamma)| = \sup_{t \ge 0} |X_0 - Y_0| \cdot \exp(-t) = |X_0 - Y_0|, \quad \forall \gamma \in \Gamma.$$

Then

$$\mathcal{M}\left\{ \lim_{|X_0 - Y_0| \to 0} \sup_{t \ge 0} |X_t - Y_t| = 0 \right\} = \mathcal{M}\left\{ \lim_{|X_0 - Y_0| \to 0} |X_0 - Y_0| = 0 \right\} = 1,$$

and the uncertain differential equation (6.66) is almost surely stable. However, the coefficient $f(t, x) = -x$ does not satisfy the strong Lipschitz condition in Theorem 6.11.

6.7 Liu's Stock Model

In 2009, Liu [38] supposed that the interest rate r_t is a constant and the stock price X_t follows an uncertain differential equation and presented an uncertain stock model with fixed interest rate as follows:

$$\begin{cases} r_t = \mu_1 \\ \mathrm{d}X_t = \mu_2 X_t \mathrm{d}t + \sigma_2 X_t \mathrm{d}C_t \end{cases} \qquad (6.67)$$

where μ_1 is the riskless interest rate, μ_2 and σ_2 are the log-drift and log-diffusion of the stock price, respectively, and C_t is a canonical Liu process. Note that the stock price is

$$X_t = X_0 \cdot \exp(\mu_2 t + \sigma_2 C_t) \qquad (6.68)$$

with an α-path

$$X_t^\alpha = X_0 \cdot \exp\left(\mu_2 t + \frac{\sqrt{3}\sigma_2 t}{\pi} \ln \frac{\alpha}{1-\alpha} \right). \qquad (6.69)$$

European Option Pricing Formulas

Consider a European call option of the stock model (6.67) with a strike price K and an expiration date T. Its price is determined by

$$f_c = \exp(-\mu_1 T) \cdot E[(X_T - K)^+].$$

Theorem 6.12 (Liu [38]) *The European call option price of the stock model (6.67) with a strike price K and an expiration date T is*

$$f_c = \exp(-\mu_1 T) \int_0^1 \left(X_T^\alpha - K \right)^+ \mathrm{d}\alpha \qquad (6.70)$$

where

$$X_T^\alpha = X_0 \cdot \exp\left(\mu_2 T + \frac{\sqrt{3}\sigma_2 T}{\pi} \ln \frac{\alpha}{1-\alpha} \right). \qquad (6.71)$$

Proof Since X_t is a contour process with an α-path represented by Eq. (6.69), it follows from Theorem 4.7 that the uncertain process $(X_t - K)^+$ is a contour process with an α-path $(X_t^\alpha - K)^+$. Then by Theorem 4.3, we have

$$E[(X_T - K)^+] = \int_0^1 (X_T^\alpha - K)^+ \mathrm{d}\alpha.$$

As a result,

$$f_c = \exp(-\mu_1 T) \int_0^1 (X_T^\alpha - K)^+ d\alpha.$$

The theorem is proved.

Consider a European put option of the stock model (6.67) with a strike price K and an expiration date T. Its price is determined by

$$f_p = \exp(-\mu_1 T) \cdot E[(K - X_T)^+].$$

Theorem 6.13 (Liu [38]) *The European put option price of the stock model (6.67) with a strike price K and an expiration date T is*

$$f_p = \exp(-\mu_1 T) \int_0^1 \left(K - X_T^\alpha\right)^+ d\alpha \qquad (6.72)$$

where

$$X_T^\alpha = X_0 \cdot \exp\left(\mu_2 T + \frac{\sqrt{3}\sigma_2 T}{\pi} \ln \frac{\alpha}{1-\alpha}\right). \qquad (6.73)$$

Proof Since X_t is a contour process with an α-path represented by Eq. (6.69), it follows from Theorem 4.7 that the uncertain process $(K - X_t)^+$ is a contour process with an α-path $(K - X_t^{1-\alpha})^+$. Then by Theorem 4.3, we have

$$E[(K - X_T)^+] = \int_0^1 (K - X_T^{1-\alpha})^+ d\alpha = \int_0^1 (K - X_T^\alpha)^+ d\alpha.$$

As a result,

$$f_p = \exp(-\mu_1 T) \int_0^1 \left(K - X_T^\alpha\right)^+ d\alpha.$$

The theorem is proved.

American Option Pricing Formulas

Consider an American call option of the stock model (6.67) with a strike price K and an expiration date T. Its price is determined by

$$f_c = E\left[\sup_{0 \le t \le T} \exp(-\mu_1 t)(X_t - K)^+\right].$$

Theorem 6.14 (Chen [4]) *The American call option price of the stock model (6.67) with a strike price K and an expiration date T is*

$$f_c = \int_0^1 \sup_{0 \le t \le T} \exp(-\mu_1 t) \left(X_t^\alpha - K\right)^+ d\alpha \qquad (6.74)$$

where

$$X_t^\alpha = X_0 \cdot \exp\left(\mu_2 t + \frac{\sqrt{3}\sigma_2 t}{\pi} \ln \frac{\alpha}{1-\alpha} \right).$$

(6.75)

Proof Since X_t is a contour process with an α-path represented by Eq. (6.69), it follows from Theorem 4.7 that the uncertain process

$$\exp(-\mu_1 t)(X_t - K)^+$$

is a contour process with an α-path

$$\exp(-\mu_1 t)\left(X_t^\alpha - K \right)^+.$$

Then by Theorem 4.4, the uncertain process

$$\sup_{0 \le t \le T} \exp(-\mu_1 t)(X_t - K)^+$$

is a contour process with an α-path

$$\sup_{0 \le t \le T} \exp(-\mu_1 t)\left(X_t^\alpha - K \right)^+.$$

As a result, we have

$$f_c = \int_0^1 \sup_{0 \le t \le T} \exp(-\mu_1 t)\left(X_t^\alpha - K \right)^+ d\alpha$$

according to Theorem 4.3. The theorem is proved.

Consider an American put option of the stock model (6.67) with a strike price K and an expiration date T. Its price is determined by

$$f_p = E\left[\sup_{0 \le t \le T} \exp(-\mu_1 t)(K - X_t)^+ \right].$$

Theorem 6.15 (Chen [4]) *The American put option price of the stock model (6.67) with a strike price K and an expiration date T is*

$$f_p = \int_0^1 \sup_{0 \le t \le T} \exp(-\mu_1 t)\left(K - X_t^\alpha \right)^+ d\alpha$$

(6.76)

where

$$X_t^\alpha = X_0 \cdot \exp\left(\mu_2 t + \frac{\sqrt{3}\sigma_2 t}{\pi} \ln \frac{\alpha}{1-\alpha} \right).$$

(6.77)

Proof Since X_t is a contour process with an α-path represented by Eq. (6.69), it follows from Theorem 4.7 that the uncertain process

$$\exp(-\mu_1 t)(K - X_t)^+$$

is a contour process with an α-path

$$\exp(-\mu_1 t) \left(K - X_t^{1-\alpha} \right)^+ .$$

Then by Theorem 4.4, the uncertain process

$$\sup_{0 \le t \le T} \exp(-\mu_1 t)(K - X_t)^+$$

is a contour process with an α-path

$$\sup_{0 \le t \le T} \exp(-\mu_1 t) \left(K - X_t^{1-\alpha} \right)^+ .$$

As a result, we have

$$f_p = \int_0^1 \sup_{0 \le t \le T} \exp(-\mu_1 t) \left(K - X_t^{1-\alpha} \right)^+ d\alpha$$

$$= \int_0^1 \sup_{0 \le t \le T} \exp(-\mu_1 t) \left(K - X_t^{\alpha} \right)^+ d\alpha$$

according to Theorem 4.3. The theorem is proved.

Asian Option Pricing Formulas

Consider an Asian call option of the stock model (6.67) with a strike price K and an expiration date T. Its price is determined by

$$f_c = \exp(-\mu_1 T) \cdot E\left[\left(\frac{1}{T} \int_0^T X_s ds - K \right)^+ \right].$$

Theorem 6.16 (Sun and Chen [63]) *The Asian call option price of the stock model (6.67) with a strike price K and an expiration date T is*

$$f_c = \exp(-\mu_1 T) \int_0^1 \left(\frac{1}{T} \int_0^T X_s^{\alpha} ds - K \right)^+ d\alpha \qquad (6.78)$$

where

$$X_s^{\alpha} = X_0 \cdot \exp\left(\mu_2 s + \frac{\sqrt{3}\sigma_2 s}{\pi} \ln \frac{\alpha}{1-\alpha} \right). \qquad (6.79)$$

Proof Since X_t is a contour process with an α-path represented by Eq. (6.69), it follows from Theorems 4.6 and 4.7 that the uncertain process

$$\left(\frac{1}{T}\int_0^T X_s ds - K\right)^+$$

is a contour process with an α-path

$$\left(\frac{1}{T}\int_0^T X_s^\alpha ds - K\right)^+.$$

As a result, we have

$$f_c = \exp(-\mu_1 T)\int_0^1 \left(\frac{1}{T}\int_0^T X_s^\alpha ds - K\right)^+ d\alpha$$

according to Theorem 4.3. The theorem is proved.

Consider an Asian put option of the stock model (6.67) with a strike price K and an expiration date T. Its price is determined by

$$f_p = \exp(-\mu_1 T)\cdot E\left[\left(K - \frac{1}{T}\int_0^T X_s ds\right)^+\right].$$

Theorem 6.17 (Sun and Chen [63]) *The Asian put option price of the stock model (6.67) with a strike price K and an expiration date T is*

$$f_p = \exp(-\mu_1 T)\int_0^1 \left(K - \frac{1}{T}\int_0^T X_s^\alpha ds\right)^+ d\alpha \qquad (6.80)$$

where

$$X_s^\alpha = X_0 \cdot \exp\left(\mu_2 s + \frac{\sqrt{3}\sigma_2 s}{\pi}\ln\frac{\alpha}{1-\alpha}\right). \qquad (6.81)$$

Proof Since X_t is a contour process with an α-path represented by Eq. (6.69), it follows from Theorems 4.6 and 4.7 that the uncertain process

$$\left(K - \frac{1}{T}\int_0^T X_s ds\right)^+$$

is a contour process with an α-path

$$\left(K - \frac{1}{T}\int_0^T X_s^{1-\alpha} ds\right)^+.$$

As a result, we have

$$f_p = \exp(-\mu_1 T) \int_0^1 \left(K - \frac{1}{T} \int_0^T X_s^{1-\alpha} ds \right)^+ d\alpha$$

$$= \exp(-\mu_1 T) \int_0^1 \left(K - \frac{1}{T} \int_0^T X_s^{\alpha} ds \right)^+ d\alpha$$

according to Theorem 4.3. The theorem is proved.

6.8 Yao's Stock Model

In 2015, Yao [85] supposed that both the interest rate r_t and the stock price X_t follow uncertain differential equations and presented an uncertain stock model with floating interest rate as follows:

$$\begin{cases} dr_t = \mu_1 r_t dt + \sigma_1 r_t dC_{1t} \\ dX_t = \mu_2 X_t dt + \sigma_2 X_t dC_{2t} \end{cases} \qquad (6.82)$$

where μ_1 and σ_1 are the log-drift and log-diffusion of the interest rate, respectively, μ_2 and σ_2 are the log-drift and log-diffusion of the stock price, respectively, and C_{1t} and C_{2t} are independent canonical Liu processes. Note that the interest rate is

$$r_t = r_0 \cdot \exp(\mu_1 t + \sigma_1 C_{1t}) \qquad (6.83)$$

with an α-path

$$r_t^{\alpha} = r_0 \cdot \exp\left(\mu_1 t + \frac{\sqrt{3}\sigma_1 t}{\pi} \ln \frac{\alpha}{1-\alpha} \right), \qquad (6.84)$$

and the stock price is

$$X_t = X_0 \cdot \exp(\mu_2 t + \sigma_2 C_t) \qquad (6.85)$$

with an α-path

$$X_t^{\alpha} = X_0 \cdot \exp\left(\mu_2 t + \frac{\sqrt{3}\sigma_2 t}{\pi} \ln \frac{\alpha}{1-\alpha} \right). \qquad (6.86)$$

European Option Pricing Formulas

Consider a European call option of the stock model (6.82) with a strike price K and an expiration date T. Its price is determined by

$$f_c = E\left[\exp\left(-\int_0^T r_s ds \right) (X_T - K)^+ \right].$$

Theorem 6.18 (Yao [85]) *The European call option price of the stock model (6.82) with a strike price K and an expiration date T is*

$$f_c = \int_0^1 \exp\left(-\int_0^T r_s^{1-\alpha} ds\right)(X_T^\alpha - K)^+ d\alpha \tag{6.87}$$

where

$$r_s^{1-\alpha} = r_0 \cdot \exp\left(\mu_1 s + \frac{\sqrt{3}\sigma_1 s}{\pi}\ln\frac{1-\alpha}{\alpha}\right) \tag{6.88}$$

and

$$X_T^\alpha = X_0 \cdot \exp\left(\mu_2 T + \frac{\sqrt{3}\sigma_2 T}{\pi}\ln\frac{\alpha}{1-\alpha}\right). \tag{6.89}$$

Proof Since r_t and X_t are contour processes with α-paths represented by Eqs. (6.84) and (6.86), respectively, it follows from Theorems 4.6 and 4.7 that the uncertain process

$$\exp\left(-\int_0^t r_s ds\right)$$

is a contour process with an α-path

$$\exp\left(-\int_0^t r_s^{1-\alpha} ds\right),$$

and the uncertain process $(X_t - K)^+$ is a contour process with an α-path $(X_t^\alpha - K)^+$. Then by Theorem 4.7, the uncertain process

$$\exp\left(-\int_0^t r_s ds\right)(X_t - K)^+$$

is also a contour process with an α-path

$$\exp\left(-\int_0^t r_s^{1-\alpha} ds\right)(X_t^\alpha - K)^+.$$

As a result, we have

$$f_c = \int_0^1 \exp\left(-\int_0^T r_s^{1-\alpha} ds\right)(X_T^\alpha - K)^+ d\alpha$$

according to Theorem 4.3. The theorem is proved.

Consider a European put option of the stock model (6.82) with a strike price K and an expiration date T. Its price is determined by

$$f_p = E\left[\exp\left(-\int_0^T r_s ds\right)(K - X_T)^+\right].$$

Theorem 6.19 (Yao [85]) *The European put option price of the stock model (6.82) with a strike price K and an expiration date T is*

$$f_p = \int_0^1 \exp\left(-\int_0^T r_s^\alpha ds\right)(K - X_T^\alpha)^+ d\alpha \tag{6.90}$$

where

$$r_s^\alpha = r_0 \cdot \exp\left(\mu_1 s + \frac{\sqrt{3}\sigma_1 s}{\pi}\ln\frac{\alpha}{1-\alpha}\right) \tag{6.91}$$

and

$$X_T^\alpha = X_0 \cdot \exp\left(\mu_2 T + \frac{\sqrt{3}\sigma_2 T}{\pi}\ln\frac{\alpha}{1-\alpha}\right). \tag{6.92}$$

Proof Since r_t and X_t are contour processes with α-paths represented by Eqs. (6.84) and (6.86), respectively, it follows from Theorems 4.6 and 4.7 that the uncertain process

$$\exp\left(-\int_0^t r_s ds\right)$$

is a contour process with an α-path

$$\exp\left(-\int_0^t r_s^{1-\alpha} ds\right),$$

and the uncertain process $(K - X_t)^+$ is a contour process with an α-path $(K - X_t^{1-\alpha})^+$. Then by Theorem 4.7, the uncertain process

$$\exp\left(-\int_0^t r_s ds\right)(K - X_t)^+$$

is also a contour process with an α-path

$$\exp\left(-\int_0^t r_s^{1-\alpha} ds\right)(K - X_t^{1-\alpha})^+.$$

As a result, we have

$$f_p = \int_0^1 \exp\left(-\int_0^T r_s^{1-\alpha} ds\right) (K - X_T^{1-\alpha})^+ d\alpha$$

$$= \int_0^1 \exp\left(-\int_0^T r_s^\alpha ds\right) (K - X_T^\alpha)^+ d\alpha$$

according to Theorem 4.3. The theorem is proved.

American Option Pricing Formulas

Consider an American call option of the stock model (6.82) with a strike price K and an expiration date T. Its price is determined by

$$f_c = E\left[\sup_{0 \le t \le T} \exp\left(-\int_0^t r_s ds\right) (X_t - K)^+\right].$$

Theorem 6.20 (Yao [85]) *The American call option price of the stock model (6.82) with a strike price K and an expiration date T is*

$$f_c = \int_0^1 \sup_{0 \le t \le T} \exp\left(-\int_0^t r_s^{1-\alpha} ds\right) (X_t^\alpha - K)^+ d\alpha \tag{6.93}$$

where

$$r_s^{1-\alpha} = r_0 \cdot \exp\left(\mu_1 s + \frac{\sqrt{3}\sigma_1 s}{\pi} \ln \frac{1-\alpha}{\alpha}\right) \tag{6.94}$$

and

$$X_t^\alpha = X_0 \cdot \exp\left(\mu_2 t + \frac{\sqrt{3}\sigma_2 t}{\pi} \ln \frac{\alpha}{1-\alpha}\right). \tag{6.95}$$

Proof It follows from Theorem 6.18 that the uncertain process

$$\exp\left(-\int_0^t r_s ds\right) (X_t - K)^+$$

is a contour process with an α-path

$$\exp\left(-\int_0^t r_s^{1-\alpha} ds\right) (X_t^\alpha - K)^+.$$

Then by Theorem 4.4, the uncertain process

$$\sup_{0 \le t \le T} \exp\left(-\int_0^t r_s ds\right) (X_t - K)^+$$

is a contour process with an α-path

$$\sup_{0 \le t \le T} \exp\left(-\int_0^t r_s^{1-\alpha} ds\right) (X_t^\alpha - K)^+.$$

As a result, we have

$$f_c = \int_0^1 \sup_{0 \le t \le T} \exp\left(-\int_0^t r_s^{1-\alpha} ds\right) (X_T^\alpha - K)^+ d\alpha$$

according to Theorem 4.3. The theorem is proved.

Consider an American put option of the stock model (6.82) with a strike price K and an expiration date T. Its price is determined by

$$f_p = E\left[\sup_{0 \le t \le T} \exp\left(-\int_0^t r_s ds\right) (K - X_t)^+\right].$$

Theorem 6.21 (Yao [85]) *The American put option price of the stock model (6.82) with a strike price K and an expiration date T is*

$$f_p = \int_0^1 \sup_{0 \le t \le T} \exp\left(-\int_0^t r_s^\alpha ds\right) (K - X_t^\alpha)^+ d\alpha \tag{6.96}$$

where

$$r_s^\alpha = r_0 \cdot \exp\left(\mu_1 s + \frac{\sqrt{3}\sigma_1 s}{\pi} \ln \frac{\alpha}{1-\alpha}\right) \tag{6.97}$$

and

$$X_t^\alpha = X_0 \cdot \exp\left(\mu_2 t + \frac{\sqrt{3}\sigma_2 t}{\pi} \ln \frac{\alpha}{1-\alpha}\right). \tag{6.98}$$

Proof It follows from Theorem 6.19 that the uncertain process

$$\exp\left(-\int_0^t r_s ds\right) (K - X_t)^+$$

is a contour process with an α-path

$$\exp\left(-\int_0^t r_s^{1-\alpha}ds\right)(K - X_t^{1-\alpha})^+.$$

Then by Theorem 4.4, the uncertain process

$$\sup_{0\le t\le T} \exp\left(-\int_0^t r_s ds\right)(K - X_t)^+$$

is a contour process with an α-path

$$\sup_{0\le t\le T} \exp\left(-\int_0^t r_s^{1-\alpha}ds\right)(K - X_t^{1-\alpha})^+.$$

As a result, we have

$$f_p = \int_0^1 \sup_{0\le t\le T} \exp\left(-\int_0^t r_s^{1-\alpha}ds\right)(K - X_t^{1-\alpha})^+ d\alpha$$

$$= \int_0^1 \sup_{0\le t\le T} \exp\left(-\int_0^t r_s^\alpha ds\right)(K - X_t^\alpha)^+ d\alpha$$

according to Theorem 4.3. The theorem is proved.

Asian Option Pricing Formulas

Consider an Asian call option of the stock model (6.82) with a strike price K and an expiration date T. Its price is determined by

$$f_c = E\left[\exp\left(-\int_0^T r_s ds\right)\cdot\left(\frac{1}{T}\int_0^T X_s ds - K\right)^+\right].$$

Theorem 6.22 (Yao [85]) *The Asian call option price of the stock model (6.82) with a strike price K and an expiration date T is*

$$f_c = \int_0^1 \exp\left(-\int_0^T r_s^{1-\alpha}ds\right)\cdot\left(\frac{1}{T}\int_0^T X_s^\alpha ds - K\right)^+ d\alpha \qquad (6.99)$$

where

$$r_s^{1-\alpha} = r_0\cdot\exp\left(\mu_1 s + \frac{\sqrt{3}\sigma_1 s}{\pi}\ln\frac{1-\alpha}{\alpha}\right) \qquad (6.100)$$

and

$$X_s^\alpha = X_0 \cdot \exp\left(\mu_2 s + \frac{\sqrt{3}\sigma_2 s}{\pi}\ln\frac{\alpha}{1-\alpha}\right).$$ (6.101)

Proof Since r_t and X_t are contour processes with α-paths represented by Eqs. (6.84) and (6.86), respectively, it follows from Theorems 4.6 and 4.7 that the uncertain process

$$\exp\left(-\int_0^t r_s ds\right)$$

is a contour process with an α-path

$$\exp\left(-\int_0^t r_s^{1-\alpha}ds\right),$$

and the uncertain process

$$\left(\frac{1}{t}\int_0^t X_s ds - K\right)^+$$

is a contour process with an α-path

$$\left(\frac{1}{t}\int_0^t X_s^\alpha ds - K\right)^+.$$

Then by Theorem 4.7, the uncertain process

$$\exp\left(-\int_0^t r_s ds\right)\cdot\left(\frac{1}{t}\int_0^t X_s ds - K\right)^+$$

is also a contour process with an α-path

$$\exp\left(-\int_0^t r_s^{1-\alpha}ds\right)\cdot\left(\frac{1}{t}\int_0^t X_s^\alpha ds - K\right)^+.$$

As a result, we have

$$f_c = \int_0^1 \exp\left(-\int_0^T r_s^{1-\alpha}ds\right)\cdot\left(\frac{1}{T}\int_0^T X_s^\alpha ds - K\right)^+ d\alpha$$

according to Theorem 4.3. The theorem is proved.

Consider an Asian put option of the stock model (6.82) with a strike price K and an expiration date T. Its price is determined by

$$f_p = E\left[\exp\left(-\int_0^T r_s ds\right) \cdot \left(K - \frac{1}{T}\int_0^T X_s ds\right)^+\right].$$

Theorem 6.23 (Yao [85]) *The Asian put option price of the stock model (6.82) with a strike price K and an expiration date T is*

$$f_p = \int_0^1 \exp\left(-\int_0^T r_s^\alpha ds\right) \cdot \left(K - \frac{1}{T}\int_0^T X_s^\alpha ds\right)^+ d\alpha \qquad (6.102)$$

where

$$r_s^\alpha = r_0 \cdot \exp\left(\mu_1 s + \frac{\sqrt{3}\sigma_1 s}{\pi}\ln\frac{\alpha}{1-\alpha}\right) \qquad (6.103)$$

and

$$X_s^\alpha = X_0 \cdot \exp\left(\mu_2 s + \frac{\sqrt{3}\sigma_2 s}{\pi}\ln\frac{\alpha}{1-\alpha}\right). \qquad (6.104)$$

Proof Since r_t and X_t are contour processes with α-paths represented by Eqs. (6.84) and (6.86), respectively, it follows from Theorems 4.6 and 4.7 that the uncertain process

$$\exp\left(-\int_0^t r_s ds\right)$$

is a contour process with an α-path

$$\exp\left(-\int_0^t r_s^{1-\alpha} ds\right),$$

and the uncertain process

$$\left(K - \frac{1}{t}\int_0^t X_s ds\right)^+$$

is a contour process with an α-path

$$\left(K - \frac{1}{t}\int_0^t X_s^{1-\alpha} ds\right)^+.$$

Then by Theorem 4.7, the uncertain process

$$\exp\left(-\int_0^t r_s ds\right) \cdot \left(K - \frac{1}{t}\int_0^t X_s ds\right)^+$$

is also a contour process with an α-path

$$\exp\left(-\int_0^t r_s^{1-\alpha}\mathrm{d}s\right) \cdot \left(K - \frac{1}{t}\int_0^t X_s^{1-\alpha}\mathrm{d}s\right)^+ .$$

As a result, we have

$$\begin{aligned}
f_p &= \int_0^1 \exp\left(-\int_0^T r_s^{1-\alpha}\mathrm{d}s\right) \cdot \left(K - \frac{1}{T}\int_0^T X_s^{1-\alpha}\mathrm{d}s\right)^+ \mathrm{d}\alpha \\
&= \int_0^1 \exp\left(-\int_0^T r_s^{\alpha}\mathrm{d}s\right) \cdot \left(K - \frac{1}{T}\int_0^T X_s^{\alpha}\mathrm{d}s\right)^+ \mathrm{d}\alpha
\end{aligned}$$

according to Theorem 4.3. The theorem is proved.

Chapter 7
Uncertain Calculus with Renewal Process

Uncertain renewal process is a nonnegative integer valued uncertain process, which counts the number of renewals that an uncertain system occurs. Uncertain calculus with renewal process deals with the integration and differentiation of uncertain processes with respect to renewal processes. The emphases of this chapter are on Yao integral and Yao process as well as the fundamental theorem and integration by parts.

7.1 Renewal Process

Definition 7.1 (*Liu [37]*) Let $\xi_1, \xi_2, \ldots$ be iid positive uncertain variables. Define $S_0 = 0$ and $S_n = \xi_1 + \xi_2 + \cdots + \xi_n$ for $n \geq 1$. Then the uncertain process

$$N_t = \max_{n \geq 0} \left\{ n \mid S_n \leq t \right\} \tag{7.1}$$

is called an uncertain renewal process.

Let $\xi_1, \xi_2, \ldots$ denote the interarrival times of successive events. Then S_n is the total waiting time before the occurrence of the nth event, and N_t is the number of renewals in $(0, t]$. The fundamental relationship formulas of an uncertain renewal process are

$$N_t \geq n \Leftrightarrow S_n \leq t,$$

$$N_t \leq n \Leftrightarrow S_{n+1} > t.$$

Theorem 7.1 (Liu [40]) *Let N_t be an uncertain renewal process. If the interarrival times have a common uncertainty distribution Φ, then N_t has an uncertainty distribution*

© Springer-Verlag Berlin Heidelberg 2016
K. Yao, *Uncertain Differential Equations*,
Springer Uncertainty Research, DOI 10.1007/978-3-662-52729-0_7

$$\Upsilon_t(x) = 1 - \Phi\left(\frac{t}{\lfloor x \rfloor + 1}\right), \quad \forall x \geq 0 \tag{7.2}$$

where $\lfloor x \rfloor$ represents the maximal integer less than or equal to x.

Proof Note that

$$\mathcal{M}\{N_t \leq n\} = \mathcal{M}\{S_{n+1} > t\} = 1 - \mathcal{M}\{S_{n+1} \leq t\} = 1 - \Phi\left(\frac{t}{n+1}\right).$$

Since N_t takes integer values, we have

$$\Upsilon_t(x) = \mathcal{M}\{N_t \leq x\} = \mathcal{M}\{N_t \leq \lfloor x \rfloor\} = 1 - \Phi\left(\frac{t}{\lfloor x \rfloor + 1}\right)$$

for any $x \geq 0$. The theorem is proved.

Theorem 7.2 *Let N_t be an uncertain renewal process. Then we have*

$$\mathcal{M}\{N_t < \infty\} = 1 \tag{7.3}$$

for any time t.

Proof Assume the uncertain interarrival times $\xi_1, \xi_2, \ldots$ have a common uncertainty distribution Φ. Then

$$\{N_t = \infty\} = \bigcap_{n=1}^{\infty}\{N_t \geq n\} \subset \{N_t \geq k\} = \{S_k \leq t\}$$

for any positive integer k. By the monotonicity of uncertain measure, we have

$$\mathcal{M}\{N_t = \infty\} \leq \mathcal{M}\{S_k \leq t\} = \mathcal{M}\{\xi_1 \leq t/k\} = \Phi(t/k).$$

Since the above inequality holds for any positive integer k, we further have

$$\mathcal{M}\{N_t = \infty\} \leq \lim_{k \to \infty} \Phi(t/k) = 0.$$

It follows from the duality of uncertain measure that

$$\mathcal{M}\{N_t < \infty\} = 1 - \mathcal{M}\{N_t = \infty\} = 1.$$

The theorem is proved.

The renewal process N_t is the number of renewals in $(0, t]$, so $\lim_{t \to \infty} N_t$ is the total number of renewals.

Theorem 7.3 *Let N_t be an uncertain renewal process. Then we have*

$$\mathcal{M}\left\{\lim_{t\to\infty} N_t = \infty\right\} = 1. \tag{7.4}$$

Proof It follows from the subadditivity of uncertain measure that

$$\mathcal{M}\left\{\lim_{t\to\infty} N_t < \infty\right\} = \mathcal{M}\left\{\bigcup_{n=1}^{\infty}(\lim_{t\to\infty} N_t < n)\right\}$$

$$\leq \sum_{n=1}^{\infty} \mathcal{M}\left\{\lim_{t\to\infty} N_t < n\right\} = \sum_{n=1}^{\infty} \mathcal{M}\{S_n = \infty\} = 0.$$

By the duality of uncertain measure, we have

$$\mathcal{M}\left\{\lim_{t\to\infty} N_t = \infty\right\} = 1 - \mathcal{M}\left\{\lim_{t\to\infty} N_t < \infty\right\} = 1.$$

The theorem is proved.

7.2 Yao Integral

Definition 7.2 (*Yao [72]*) Let X_t be an uncertain process and N_t be an uncertain renewal process. Then the Yao integral of X_t with respect to N_t on the interval $[a, b]$ is defined by

$$\int_a^b X_t dN_t = \sum_{a < t \leq b} X_{t-}(N_t - N_{t-}) \tag{7.5}$$

provided that the sum exists almost surely and is finite. In this case, the uncertain process X_t is said to be Yao integrable.

Remark 7.1 Assume that X_t is a Yao integrable uncertain process on an uncertainty space $(\Gamma, \mathcal{L}, \mathcal{M})$. Then for each $\gamma \in \Gamma$, the integral

$$\int_0^t X_s(\gamma) dN_s(\gamma)$$

is a sample path of

$$\int_0^t X_s dN_s.$$

Example 7.1 Consider the uncertain renewal process N_t. Since

$$N_t = \sum_{0 < s \le t} (N_s - N_{s-}),$$

we have

$$\int_0^t dN_s = N_t.$$

Example 7.2 Consider the uncertain process N_t^2. Since

$$N_t^2 = \sum_{0 < s \le t} \left(N_{s_{i+1}}^2 - N_{s_i}^2 \right)$$

$$= \sum_{0 < s \le t} \left(N_{s_{i+1}} - N_{s_i} \right)^2 + 2 \sum_{0 < s \le t} N_{s_i} \left(N_{s_{i+1}} - N_{s_i} \right)$$

$$= N_t + 2 \int_0^t N_{s-} dN_s,$$

we have

$$\int_0^t N_{s-} dN_s = \frac{1}{2} N_t (N_t - 1).$$

Example 7.3 For any partition $0 = s_1 < s_2 < \cdots < s_{k+1} = t$, since

$$t N_t = \sum_{i=1}^k \left(s_{i+1} N_{s_{i+1}} - s_i N_{s_i} \right)$$

$$= \sum_{i=1}^k N_{s_{i+1}} \left(s_{i+1} - s_i \right) + \sum_{i=1}^k s_i \left(N_{s_{i+1}} - N_{s_i} \right)$$

$$\to \int_0^t N_s ds + \int_0^t s dN_s$$

as $\Delta \to 0$, we have

$$\int_0^t N_s ds + \int_0^t s dN_s = t N_t.$$

Theorem 7.4 (Yao [72]) *If X_t is a Yao integrable uncertain process on $[a, b]$, then it is Yao integrable on each subinterval of $[a, b]$. Moreover, if $c \in [a, b]$, then*

$$\int_a^b X_t dN_t = \int_a^c X_t dN_t + \int_c^b X_t dN_t. \tag{7.6}$$

Proof Let $[a', b']$ be a subinterval of $[a, b]$. Since X_t is a Yao integrable uncertain process on $[a, b]$, the sum

$$\sum_{a < t \leq b} X_{t-}(N_t - N_{t-})$$

exists almost surely and is finite. So the sum

$$\sum_{a' < t \leq b'} X_{t-}(N_t - N_{t-})$$

also exists almost surely and is finite, and the uncertain process X_t is Yao integrable on $[a', b']$. Next, for a real number $c \in [a, b]$, we have

$$\int_a^b X_t dN_t = \sum_{a < t \leq b} X_{t-}(N_t - N_{t-})$$

$$= \sum_{a < t \leq c} X_{t-}(N_t - N_{t-}) + \sum_{c < t \leq b} X_{t-}(N_t - N_{t-})$$

$$= \int_a^c X_t dN_t + \int_b^c X_t dN_t.$$

The theorem is proved.

Theorem 7.5 (Yao [72], Linearity of Yao Integral) *Let X_t and Y_t be two Yao integrable uncertain processes on $[a, b]$. Then*

$$\int_a^b (\alpha X_t + \beta Y_t) dN_t = \alpha \int_a^b X_t dN_t + \beta \int_a^b Y_t dN_t \qquad (7.7)$$

for any real numbers α and β.

Proof It follows from Definition 7.2 of Yao integral that

$$\int_a^b (\alpha X_t + \beta Y_t) dN_t = \sum_{a < t \leq b} (\alpha X_{t-} + \beta Y_{t-})(N_t - N_{t-})$$

$$= \alpha \sum_{a < t \leq b} X_{t-}(N_t - N_{t-}) + \beta \sum_{a < t \leq b} Y_{t-}(N_t - N_{t-})$$

$$= \alpha \int_a^b X_t dN_t + \beta \int_a^b Y_t dN_t.$$

The theorem is proved.

7.3 Yao Process

Definition 7.3 (*Yang et al. [70]*) Let N_t be an uncertain renewal process, C_t be a canonical Liu process, and μ_t, σ_t, and ν_t be time integrable, Liu integrable, and Yao integrable uncertain processes, respectively. Then the uncertain process

$$Z_t = Z_0 + \int_0^t \mu_s ds + \int_0^t \sigma_s dC_s + \int_0^t \nu_s dN_s$$

is called a Yao process, and it has a Yao differential

$$dZ_t = \mu_t dt + \sigma_t dC_t + \nu_t dN_t.$$

Example 7.4 Since the uncertain renewal process N_t satisfies

$$N_t = \int_0^t dN_s,$$

it is a Yao process and has a Yao differential dN_t.

Example 7.5 Since the uncertain process $Z_t = \mu t + \sigma C_t + \nu N_t$ satisfies

$$Z_t = \int_0^t \mu ds + \int_0^t \sigma dC_t + \int_0^t \nu dN_s,$$

it is a Yao process and has a Yao differential

$$dZ_t = \mu dt + \sigma dC_t + \nu dN_t.$$

Example 7.6 Since the uncertain process N_t^2 satisfies

$$N_t^2 = 2 \int_0^t N_{s-} dN_s + \int_0^t dN_s,$$

it is a Yao process and has a Yao differential

$$dN_t^2 = 2N_{t-} dN_t + dN_t.$$

Example 7.7 Since the uncertain process $t N_t$ satisfies

$$t N_t = \int_0^t N_s ds + \int_0^t s dN_s,$$

it is a Yao process and has a Yao differential

$$d(tN_t) = N_t dt + t dN_t.$$

Fundamental Theorem

Theorem 7.6 (Yao [72], Fundamental Theorem) *Let C_t be a canonical Liu process, N_t be an uncertain renewal process, and $h(t,c,n)$ be a continuously differentiable function. Then the uncertain process $Z_t = h(t, C_t, N_t)$ is a Yao process, and it has a Yao differential*

$$dZ_t = \frac{\partial h}{\partial t}(t, C_t, N_t)dt + \frac{\partial h}{\partial c}(t, C_t, N_t)dC_t + h(t, C_t, N_t) - h(t, C_t, N_{t-}). \quad (7.8)$$

Proof Note that $\Delta C_t = C_t - C_{t-\Delta t}$ is an infinitesimal with the same order as Δt. By using Taylor series expansion, we get a first-order approximation

$$\Delta Z_t = \frac{\partial h}{\partial t}(t, C_t, N_t)\Delta t + \frac{\partial h}{\partial c}(t, C_t, N_t)\Delta C_t + h(t, C_t, N_t) - h(t, C_t, N_{t-\Delta t}).$$

Letting $\Delta t \to 0$, we have

$$dZ_t = \frac{\partial h}{\partial t}(t, C_t, N_t)dt + \frac{\partial h}{\partial C_t}(t, C_t, N_t)dC_t + h(t, C_t, N_t) - h(t, C_t, N_{t-}).$$

The theorem is proved.

Example 7.8 Consider the Yao differential of the uncertain process $\mu t + \sigma C_t + \nu N_t$. In this case, we assume $h(t, c, n) = \mu t + \sigma c + \nu n$. Since

$$\frac{\partial h}{\partial t}(t, c, n) = \mu, \quad \frac{\partial h}{\partial c}(t, c, n) = \sigma, \quad h(t, C_t, N_t) - h(t, C_t, N_{t-}) = \nu dN_t,$$

we have

$$d(\mu t + \sigma C_t + \gamma N_t) = \mu dt + \sigma dC_t + \nu dN_t.$$

Example 7.9 Consider the Yao differential of the uncertain process $tC_t N_t$. In this case, we assume $h(t, c, n) = tcn$. Since

$$\frac{\partial h}{\partial t}(t, c, n) = cn, \quad \frac{\partial h}{\partial c}(t, c, n) = tn, \quad h(t, C_t, N_t) - h(t, C_t, N_{t-}) = tC_t dN_t,$$

we have

$$d(tC_t N_t) = C_t N_t dt + t N_t dC_t + t C_t dN_t.$$

Theorem 7.7 (Yang et al. [70]) *Let X_t be a Yao process such that $dX_t = \mu_t dt + \sigma_t dC_t + \nu_t dN_t$, and $h(t, x)$ be a continuously differentiable function. Then the uncertain process $Z_t = h(t, X_t)$ is also a Yao process, and it has a Yao differential*

$$dZ_t = \left(\frac{\partial h}{\partial t}(t, X_t) + \mu_t \cdot \frac{\partial h}{\partial x}(t, X_t) \right) dt + \sigma_t \cdot \frac{\partial h}{\partial x}(t, X_t) dC_t$$
$$+ (h(t, X_{t-} + \nu_t) - h(t, X_{t-})) dN_t. \qquad (7.9)$$

Proof By using Taylor series expansion, we get a first-order approximation

$$\Delta Z_t = \frac{\partial h}{\partial t}(t, X_t) \Delta t + h(t, X_t) - h(t, X_{t-\Delta t}).$$

Since $\Delta C_t = C_t - C_{t-\Delta t}$ is an infinitesimal with the same order as Δt, we have a first-order approximation

$$h(t, X_t) - h(t, X_{t-\Delta t}) = \frac{\partial h}{\partial x}(t, X_t)(\mu_t \Delta t + \sigma_t \Delta C_t)$$
$$+ h\left(t, X_{t-\Delta t} + \int_{t-\Delta t}^{t} \nu_s dN_s \right) - h(t, X_{t-\Delta t}).$$

Letting $\Delta t \to 0$, we have

$$dZ_t = \frac{\partial h}{\partial t}(t, X_t) dt + h(t, X_t) - h(t, X_{t-})$$

and

$$h(t, X_t) - h(t, X_{t-}) = \frac{\partial h}{\partial x}(t, X_t)(\mu_t dt + \sigma_t dC_t)$$
$$+ (h(t, X_{t-} + \nu_t) - h(t, X_{t-})) dN_t.$$

As a result,

$$dZ_t = \left(\frac{\partial h}{\partial t}(t, X_t) + \mu_t \cdot \frac{\partial h}{\partial x}(t, X_t) \right) dt + \sigma_t \cdot \frac{\partial h}{\partial x}(t, X_t) dC_t$$
$$+ (h(t, X_{t-} + \nu_t) - h(t, X_{t-})) dN_t.$$

The theorem is proved.

Integration by Parts

Theorem 7.8 (Yao [72], Integration by Parts) *Let X_t and Y_t be two Yao processes. Then*

$$d(X_t Y_t) = Y_t dX_t + X_t dY_t + (X_t - X_{t-})(Y_t - Y_{t-}).$$

Proof For any partition of closed interval of $[0, t]$ with $0 = t_1 < t_2 < \cdots < t_{k+1} = t$, we have

$$X_t Y_t = X_0 Y_0 + \lim_{\Delta \to 0} \sum_{i=1}^{k} (X_{t_{i+1}} Y_{t_{i+1}} - X_{t_i} Y_{t_i})$$

$$= X_0 Y_0 + \lim_{\Delta \to 0} \sum_{i=1}^{k} \left(X_{t_i}(Y_{t_{i+1}} - Y_{t_i}) + Y_{t_i}(X_{t_{i+1}} - X_{t_i}) \right.$$

$$\left. + (X_{t_{i+1}} - X_{t_i})(Y_{t_{i+1}} - Y_{t_i}) \right).$$

Then we have

$$X_t Y_t = X_0 Y_0 + \int_0^t Y_s dX_s + \int_0^t X_s dY_s + \sum_{0 < s \le t} (X_s - X_{s-})(Y_s - Y_{s-}),$$

which is equivalent to

$$d(X_t Y_t) = Y_t dX_t + X_t dY_t + (X_t - X_{t-})(Y_t - Y_{t-}).$$

The theorem is proved.

Chapter 8
Uncertain Differential Equation with Jumps

Uncertain differential equation with jumps is a type of differential equations driven by both canonical Liu processes and uncertain renewal processes. This chapter introduces uncertain differential equation with jumps, including existence and uniqueness theorem, and stability theorems as well as its application in stock markets.

8.1 Uncertain Differential Equation with Jumps

Definition 8.1 (*Yao [72]*) Suppose that C_t is a canonical Liu process, N_t is an uncertain renewal process, and f, g, and h are some measurable functions. Then

$$dX_t = f(t, X_t)dt + g(t, X_t)dC_t + h(t, X_t)dN_t \tag{8.1}$$

is called an uncertain differential equation with jumps. An uncertain process that satisfies (8.1) identically at each time t is called a solution of the uncertain differential equation with jumps.

Remark 8.1 The uncertain differential equation with jumps (8.1) is equivalent to the uncertain integral equation

$$X_t = X_0 + \int_0^t f(s, X_s)ds + \int_0^t g(s, X_s)dC_s + \int_0^t h(s, X_s)dN_s.$$

Apparently, the solution of an uncertain differential equation with jumps is a Yao process.

Theorem 8.1 *Let C_t be a canonical Liu process, N_t be an uncertain renewal process with iid interarrival times $\xi_1, \xi_2, \ldots,$ and μ_t, σ_t, and ν_t be some measurable functions. Then the uncertain differential equation with jumps*

$$dX_t = \mu_t dt + \sigma_t dC_t + \nu_t dN_t \tag{8.2}$$

© Springer-Verlag Berlin Heidelberg 2016
K. Yao, *Uncertain Differential Equations*,
Springer Uncertainty Research, DOI 10.1007/978-3-662-52729-0_8

has a solution

$$X_t = X_0 + \int_0^t \mu_s ds + \int_0^t \sigma_s dC_s + \sum_{i=1}^{N_t} \nu_{S_i} \qquad (8.3)$$

where $S_0 = 0$ and $S_i = \xi_1 + \xi_2 + \cdots + \xi_i$ for $i \geq 1$.

Proof We will prove the theorem by showing that the differential equation

$$dX_t(\gamma) = \mu_t dt + \sigma_t dC_t(\gamma) + \nu_t dN_t(\gamma), \quad X_0(\gamma) = X_0 \qquad (8.4)$$

has a solution

$$X_t(\gamma) = X_0 + \int_0^t \mu_s ds + \int_0^t \sigma_s dC_s(\gamma) + \sum_{i=1}^{N_t(\gamma)} \nu_{S_i(\gamma)}$$

for almost all γ. Given any time u, consider the solution of (8.4) on the interval $[0, u]$. Without loss of generality, we assume $N_u(\gamma) = n$, i.e., $S_n(\gamma) \leq u < S_{n+1}(\gamma)$. Since $dN_t(\gamma) = 0$ for any $t \in [0, S_1(\gamma))$, the differential equation (8.4) degenerates to

$$dX_t(\gamma) = \mu_t dt + \sigma_t dC_t(\gamma), \quad X_0(\gamma) = X_0$$

whose solution is

$$X_t(\gamma) = X_0 + \int_0^t \mu_s ds + \int_0^t \sigma_s dC_s(\gamma).$$

Since a jump occurs at the time $S_1(\gamma)$, we have

$$X_{S_1(\gamma)}(\gamma) = \lim_{t \to (S_1(\gamma))^-} X_t(\gamma) + \nu_{S_1(\gamma)}$$

$$= X_0 + \int_0^{S_1(\gamma)} \mu_s ds + \int_0^{S_1(\gamma)} \sigma_s dC_s(\gamma) + \nu_{S_1(\gamma)}.$$

Since $dN_t(\gamma) = 0$ for any $t \in (S_1(\gamma), S_2(\gamma))$, the differential equation (8.4) degenerates to

$$dX_t(\gamma) = \mu_t dt + \sigma_t dC_t(\gamma)$$

with an initial value $X_{S_1(\gamma)}(\gamma)$, whose solution is

$$X_t(\gamma) = X_{S_1(\gamma)}(\gamma) + \int_{S_1(\gamma)}^t \mu_s ds + \int_{S_1(\gamma)}^t \sigma_s dC_s(\gamma)$$

$$= X_0 + \int_0^t \mu_s ds + \int_0^t \sigma_s dC_s(\gamma) + \nu_{S_1(\gamma)}.$$

Since a jump occurs at the time $S_2(\gamma)$, we have

$$X_{S_2(\gamma)}(\gamma) = \lim_{t \to (S_2(\gamma))^-} X_t(\gamma) + \nu_{S_2(\gamma)}$$

$$= X_0 + \int_0^{S_2(\gamma)} \mu_s ds + \int_0^{S_2(\gamma)} \sigma_s dC_s(\gamma) + \nu_{S_1(\gamma)} + \nu_{S_2(\gamma)}.$$

Repeating the above operations, we obtain that

$$X_{S_n(\gamma)}(\gamma) = X_0 + \int_0^{S_n(\gamma)} \mu_s ds + \int_0^{S_n(\gamma)} \sigma_s dC_s(\gamma) + \sum_{i=1}^{n} \nu_{S_i(\gamma)}.$$

Since $dN_t(\gamma) = 0$ on the interval $(S_n(\gamma), u]$, the differential equation (8.4) degenerates to

$$dX_t(\gamma) = \mu_t dt + \sigma_t dC_t(\gamma)$$

with an initial value $X_{S_n}(\gamma)$, whose solution is

$$X_u(\gamma) = X_{S_n(\gamma)}(\gamma) + \int_{S_n(\gamma)}^{u} \mu_s ds + \int_{S_n(\gamma)}^{u} \sigma_s dC_s(\gamma)$$

$$= X_0 + \int_0^{u} \mu_s ds + \int_0^{u} \sigma_s dC_s(\gamma) + \sum_{i=1}^{n} \nu_{S_i(\gamma)}$$

$$= X_0 + \int_0^{u} \mu_s ds + \int_0^{u} \sigma_s dC_s(\gamma) + \sum_{i=1}^{N_u(\gamma)} \nu_{S_i(\gamma)}.$$

The theorem is proved.

Example 8.1 Assume that μ, σ, and ν are three real numbers. Then the uncertain differential equation with jumps

$$dX_t = \mu dt + \sigma dC_t + \nu dN_t$$

has a solution

$$X_t = X_0 + \mu t + \sigma C_t + \nu N_t.$$

Theorem 8.2 *Let C_t be a canonical Liu process, N_t be an uncertain renewal process with iid interarrival times $\xi_1, \xi_2, \ldots$, and μ_t, σ_t, and ν_t be some measurable functions. Then the uncertain differential equation with jumps*

$$dX_t = \mu_t X_t dt + \sigma_t X_t dC_t + \nu_t X_t dN_t \tag{8.5}$$

has a solution

$$X_t = X_0 \cdot \exp\left(\int_0^t \mu_s ds + \int_0^t \sigma_s dC_s\right) \cdot \prod_{i=1}^{N_t}(1 + \nu_{S_i}) \tag{8.6}$$

where $S_0 = 0$ and $S_i = \xi_1 + \xi_2 + \cdots + \xi_i$ for $i \geq 1$.

Proof We will prove the theorem by showing that the differential equation

$$dX_t(\gamma) = \mu_t X_t(\gamma)dt + \sigma_t X_t(\gamma)dC_t(\gamma) + \nu_t X_t(\gamma)dN_t(\gamma), \quad X_0(\gamma) = X_0 \tag{8.7}$$

has a solution

$$X_t(\gamma) = X_0 \cdot \exp\left(\int_0^t \mu_s ds + \int_0^t \sigma_s dC_s(\gamma)\right) \cdot \prod_{i=1}^{N_t(\gamma)}\left(1 + \nu_{S_i(\gamma)}\right)$$

for almost all γ. Given any time u, consider the solution of (8.7) on the interval $[0, u]$. Without loss of generality, we assume $N_u(\gamma) = n$, i.e., $S_n(\gamma) \leq u < S_{n+1}(\gamma)$. Since $dN_t(\gamma) = 0$ for any $t \in [0, S_1(\gamma))$, the differential equation (8.7) degenerates to

$$dX_t(\gamma) = \mu_t X_t(\gamma)dt + \sigma_t X_t(\gamma)dC_t(\gamma), \quad X_0(\gamma) = X_0$$

whose solution is

$$X_t(\gamma) = X_0 \cdot \exp\left(\int_0^t \mu_s ds + \int_0^t \sigma_s dC_s(\gamma)\right).$$

Since a jump occurs at the time $S_1(\gamma)$, we have

$$X_{S_1(\gamma)}(\gamma) = \lim_{t \to (S_1(\gamma))^-} X_t(\gamma) + \nu_{S_1(\gamma)} \cdot \lim_{t \to (S_1(\gamma))^-} X_t(\gamma)$$
$$= X_0 \cdot \exp\left(\int_0^{S_1(\gamma)} \mu_s ds + \int_0^{S_1(\gamma)} \sigma_s dC_s(\gamma)\right) \cdot \left(1 + \nu_{S_1(\gamma)}\right).$$

Since $dN_t(\gamma) = 0$ for any $t \in (S_1(\gamma), S_2(\gamma))$, the differential equation (8.7) degenerates to

$$dX_t(\gamma) = \mu_t X_t(\gamma)dt + \sigma_t X_t(\gamma)dC_t(\gamma)$$

with an initial value $X_{S_1(\gamma)}(\gamma)$, whose solution is

$$X_t(\gamma) = X_{S_1(\gamma)}(\gamma) \cdot \exp\left(\int_{S_1(\gamma)}^t \mu_s ds + \int_{S_1(\gamma)}^t \sigma_s dC_s(\gamma)\right)$$
$$= X_0 \cdot \exp\left(\int_0^t \mu_s ds + \int_0^t \sigma_s dC_s(\gamma)\right) \cdot \left(1 + \nu_{S_1(\gamma)}\right).$$

Since a jump occurs at the time $S_2(\gamma)$, we have

$$X_{S_2(\gamma)}(\gamma) = \lim_{t \to (S_2(\gamma))^-} X_t(\gamma) + \nu_{S_2(\gamma)} \cdot \lim_{t \to (S_2(\gamma))^-} X_t(\gamma)$$

$$= X_0 \cdot \exp\left(\int_0^{S_2(\gamma)} \mu_s ds + \int_0^{S_2(\gamma)} \sigma_s dC_s(\gamma)\right) \cdot \prod_{i=1}^{2} \left(1 + \nu_{S_i(\gamma)}\right).$$

Repeating the above operations, we obtain that

$$X_{S_n(\gamma)}(\gamma) = X_0 \cdot \exp\left(\int_0^{S_n(\gamma)} \mu_s ds + \int_0^{S_n(\gamma)} \sigma_s dC_s(\gamma)\right) \cdot \prod_{i=1}^{n} \left(1 + \nu_{S_i(\gamma)}\right).$$

Since $dN_t(\gamma) = 0$ on the interval $(S_n(\gamma), u]$, the differential equation (8.7) degenerates to

$$dX_t(\gamma) = \mu_t X_t(\gamma) dt + \sigma_t X_t(\gamma) dC_t(\gamma)$$

with an initial value $X_{S_n}(\gamma)$, whose solution is

$$X_u(\gamma) = X_{S_n(\gamma)}(\gamma) \cdot \exp\left(\int_{S_n(\gamma)}^{u} \mu_s ds + \int_{S_n(\gamma)}^{u} \sigma_s dC_s(\gamma)\right)$$

$$= X_0 \cdot \exp\left(\int_0^{u} \mu_s ds + \int_0^{u} \sigma_s dC_s(\gamma)\right) \cdot \prod_{i=1}^{n} \left(1 + \nu_{S_i(\gamma)}\right)$$

$$= X_0 \cdot \exp\left(\int_0^{u} \mu_s ds + \int_0^{u} \sigma_s dC_s(\gamma)\right) \cdot \prod_{i=1}^{N_u(\gamma)} \left(1 + \nu_{S_i(\gamma)}\right).$$

The theorem is proved.

Example 8.2 Assume that μ, σ, and ν are three real numbers. Then the uncertain differential equation with jumps

$$dX_t = \mu X_t dt + \sigma X_t dC_t + \nu X_t dN_t$$

has a solution

$$X_t = X_0 \cdot \exp(\mu t + \sigma C_t) \cdot (1 + \nu)^{N_t}.$$

8.2 Existence and Uniqueness Theorem

In this section, we give a sufficient condition for an uncertain differential equation with jumps having a unique solution.

Theorem 8.3 (Yao [83]) *The uncertain differential equation with jumps*

$$dX_t = f(t, X_t) dt + g(t, X_t) dC_t + h(t, X_t) dN_t \tag{8.8}$$

has a unique solution if the coefficients $f(t, x)$ and $g(t, x)$ satisfy the linear growth condition

$$|f(t, x)| + |g(t, x)| \leq L(1 + |x|), \quad \forall x \in \Re, t \geq 0 \tag{8.9}$$

and the Lipschitz condition

$$|f(t, x) - f(t, y)| + |g(t, x) - g(t, y)| \leq L|x - y|, \quad \forall x, y \in \Re, t \geq 0 \tag{8.10}$$

for some constants L, and the coefficient $h(t, x)$ is continuous.

Proof For a Lipschitz continuous sample path $C_t(\gamma)$, consider the differential equation

$$dX_t(\gamma) = f(t, X_t(\gamma))dt + g(t, X_t(\gamma))dC_t(\gamma) + h(t, X_t(\gamma))dN_t(\gamma) \tag{8.11}$$

with an initial value $X_0(\gamma) = X_0$. Given any time T, we will prove that the differential equation (8.11) has a unique solution on the interval $[0, T]$. Let $\xi_1, \xi_2, \ldots$ denote the interarrival times of the uncertain renewal process N_t. Write $S_0 = 0$ and $S_i = \xi_1 + \xi_2 + \ldots + \xi_i$ for $i \geq 1$. Without loss of generality, we assume $N_T(\gamma) = n$, i.e., $S_n(\gamma) \leq T < S_{n+1}(\gamma)$. Since $dN_t(\gamma) = 0$ for any $t \in [0, S_1(\gamma))$, the differential equation (8.11) degenerates to

$$dX_t(\gamma) = f(t, X_t(\gamma))dt + g(t, X_t(\gamma))dC_t(\gamma), \quad X_0(\gamma) = X_0$$

which has a unique and continuous solution in $[0, S_1(\gamma))$ by Theorem 6.7. So the limit

$$\lim_{t \to (S_1(\gamma))^-} X_t(\gamma)$$

exists, and

$$X_{S_1(\gamma)}(\gamma) = \lim_{t \to (S_1(\gamma))^-} X_t(\gamma) + h\left(S_1(\gamma), \lim_{t \to (S_1(\gamma))^-} X_t(\gamma)\right).$$

Since $dN_t(\gamma) = 0$ on the interval $(S_1(\gamma), S_2(\gamma))$, the differential equation (8.11) degenerates to

$$dX_t(\gamma) = f(t, X_t(\gamma))dt + g(t, X_t(\gamma))dC_t(\gamma)$$

with an initial value $X_{S_1(\gamma)}(\gamma)$, which has a unique and continuous solution in $(S_1(\gamma), S_2(\gamma))$ by Theorem 6.7. So the limit

$$\lim_{t \to (S_2(\gamma))^-} X_t(\gamma)$$

exists, and

$$X_{S_2(\gamma)}(\gamma) = \lim_{t \to (S_2(\gamma))^-} X_t(\gamma) + h\left(S_2(\gamma), \lim_{t \to (S_2(\gamma))^-} X_t(\gamma)\right).$$

Repeating the above operations, we obtain that $dN_t(\gamma) = 0$ on the interval $(S_n(\gamma), T]$, and the differential equation (8.11) degenerates to

$$dX_t(\gamma) = f(t, X_t(\gamma))dt + g(t, X_t(\gamma))dC_t(\gamma)$$

with an initial value $X_{S_n(\gamma)}(\gamma)$, which has a unique solution on the interval $(S_n(\gamma), T]$ by Theorem 6.7. Thus we prove that the differential equation (8.11) has a unique solution on the interval $[0, T]$. Since almost all the sample paths of a canonical Liu process C_t are Lipschitz continuous, the uncertain differential equation with jumps (8.8) has a unique solution. The theorem is proved.

8.3 Stability Theorems

In this section, we introduce the concepts of stability in two senses, namely stability in measure and almost sure stability.

Stability in Measure

Definition 8.2 (*Yao [83]*) An uncertain differential equation with jumps

$$dX_t = f(t, X_t)dt + g(t, X_t)dC_t + h(t, X_t)dN_t \tag{8.12}$$

is said to be stable in measure if for any two solutions X_t and Y_t with different initial values X_0 and Y_0, we have

$$\lim_{|X_0 - Y_0| \to 0} \mathcal{M}\left\{\sup_{t \geq 0} |X_t - Y_t| \leq \varepsilon\right\} = 1 \tag{8.13}$$

for any given number $\varepsilon > 0$.

Example 8.3 Consider an uncertain differential equation with jumps

$$dX_t = \mu dt + \sigma dC_t + \nu dN_t. \tag{8.14}$$

Since its solutions with different initial values X_0 and Y_0 are

$$X_t = X_0 + \mu t + \sigma C_t + \nu N_t,$$

$$Y_t = Y_0 + \mu t + \sigma C_t + \nu N_t,$$

respectively, we have

$$\sup_{t \geq 0} |X_t - Y_t| = |X_0 - Y_0|$$

almost surely. Then

$$\lim_{|X_0-Y_0|\to 0} \mathcal{M}\left\{\sup_{t\geq 0}|X_t - Y_t| \leq \varepsilon\right\} = \lim_{|X_0-Y_0|\to 0} \mathcal{M}\{|X_0 - Y_0| \leq \varepsilon\} = 1,$$

and the uncertain differential equation with jumps (8.14) is stable in measure.

Example 8.4 Consider an uncertain differential equation with jumps

$$dX_t = \mu X_t dt + \sigma X_t dC_t + \nu X_t dN_t \tag{8.15}$$

where μ, σ, ν are positive real numbers. Since its solutions with different initial values X_0 and Y_0 are

$$X_t = X_0 \cdot \exp(\mu t + \sigma C_t) \cdot (1 + \nu)^{N_t},$$

$$Y_t = Y_0 \cdot \exp(\mu t + \sigma C_t) \cdot (1 + \nu)^{N_t},$$

respectively, we have

$$\sup_{t\geq 0}|X_t(\gamma) - Y_t(\gamma)| = \sup_{t\geq 0}|X_0 - Y_0| \cdot \exp(\mu t + \sigma C_t(\gamma)) \cdot (1 + \nu)^{N_t(\gamma)} = +\infty$$

when $C_t(\gamma) > 0$ for all $t \in \mathfrak{R}$. Then

$$\mathcal{M}\left\{\sup_{t\geq 0}|X_t - Y_t| \leq \varepsilon\right\} = 1 - \mathcal{M}\left\{\sup_{t\geq 0}|X_t - Y_t| > \varepsilon\right\}$$

$$\leq 1 - \mathcal{M}\{C_t > 0, \forall t\} = 1/2,$$

and the uncertain differential equation with jumps (8.15) is not stable in measure.

Theorem 8.4 (Yao [83]) *Assume the uncertain differential equation with jumps*

$$dX_t = f(t, X_t)dt + g(t, X_t)dC_t + h(t, X_t)dN_t \tag{8.16}$$

has a unique solution for each given initial value. Then it is stable in measure if the coefficients $f(t, x)$ and $g(t, x)$ satisfy

$$|f(t, x) - f(t, y)| + |g(t, x) - g(t, y)| \leq L_1(t)|x - y|, \quad \forall x, y \in \mathfrak{R}, t \geq 0 \tag{8.17}$$

for some integrable function $L_1(t)$ on $[0, +\infty)$, and the coefficient $h(t, x)$ satisfies

$$|h(t, x) - h(t, y)| \leq L_2(t)|x - y|, \quad \forall x, y \in \mathfrak{R}, t \geq 0 \tag{8.18}$$

for some monotone and integrable function $L_2(t)$ on $[0, +\infty)$.

Proof Let X_t and Y_t be the solutions of the uncertain differential equation with jumps (8.16) with different initial values X_0 and Y_0, respectively. Then for a Lipschitz continuous sample path $C_t(\gamma)$, we have

$$X_t(\gamma) = X_0 + \int_0^t f(s, X_s(\gamma))ds + \int_0^t g(s, X_s(\gamma))dC_s(\gamma) + \int_0^t h(s, X_s(\gamma))dN_s(\gamma),$$

$$Y_t(\gamma) = Y_0 + \int_0^t f(s, Y_s(\gamma))ds + \int_0^t g(s, Y_s(\gamma))dC_s(\gamma) + \int_0^t h(s, Y_s(\gamma))dN_s(\gamma).$$

By the strong Lipschitz condition, we have

$$|X_t(\gamma) - Y_t(\gamma)|$$
$$\leq |X_0 - Y_0| + \int_0^t |f(s, X_s(\gamma)) - f(s, Y_s(\gamma))|ds$$
$$+ \int_0^t |g(s, X_s(\gamma)) - g(s, Y_s(\gamma))||dC_s(\gamma)| + \int_0^t |h(s, X_s(\gamma)) - h(s, Y_s(\gamma))|dN_s(\gamma)$$
$$\leq |X_0 - Y_0| + \int_0^t L_1(s)|X_s(\gamma) - Y_s(\gamma)|ds$$
$$+ \int_0^t L_1(s)K(\gamma)|X_s(\gamma) - Y_s(\gamma)|ds + \int_0^t L_2(s)|X_s(\gamma) - Y_s(\gamma)|dN_s(\gamma)$$
$$= |X_0 - Y_0| + \int_0^t (1 + K(\gamma))L_1(s)|X_s(\gamma) - Y_s(\gamma)|ds$$
$$+ \int_0^t L_2(s)|X_s(\gamma) - Y_s(\gamma)|dN_s(\gamma)$$

where $K(\gamma)$ is the Lipschitz constant of $C_t(\gamma)$. Let $\xi_1, \xi_2, \ldots$ denote the interarrival times of the uncertain renewal process N_t. Write $S_0 = 0$ and $S_i = \xi_1 + \xi_2 + \cdots + \xi_i$ for $i \geq 1$. Then

$$|X_t(\gamma) - Y_t(\gamma)|$$
$$\leq |X_0 - Y_0| \cdot \exp\left((1 + K(\gamma)) \int_0^t L_1(s)ds\right) \cdot \prod_{i=1}^{N_t(\gamma)} (1 + L_2(S_i(\gamma)))$$
$$\leq |X_0 - Y_0| \cdot \exp\left((1 + K(\gamma)) \int_0^{+\infty} L_1(s)ds\right) \cdot \prod_{i=1}^{\infty} (1 + L_2(S_i(\gamma)))$$
$$\leq |X_0 - Y_0| \cdot \exp\left((1 + K(\gamma)) \int_0^{+\infty} L_1(s)ds + \sum_{i=1}^{\infty} L_2(S_i(\gamma))\right)$$

for any $t \geq 0$. Thus we have

$$\sup_{t\geq 0}|X_t - Y_t| \leq |X_0 - Y_0| \cdot \exp\left((1 + K)\int_0^{+\infty} L_1(s)ds + \sum_{i=1}^{\infty} L_2(S_i)\right)$$

almost surely, where K is a nonnegative uncertain variable such that

$$\lim_{x\to\infty} \mathcal{M}\{\gamma \in \Gamma \mid K(\gamma) \leq x\} = 1$$

by Theorem 5.3. For any given number $\epsilon > 0$, there exists a real number H_1 such that the uncertain event $\Lambda_1 = \{\gamma \mid K(\gamma) \leq H_1\}$ has an uncertain measure $\mathcal{M}\{\Lambda_1\} \geq 1 - \epsilon$, and there exists a real number H_2 such that the uncertain event

$$\Lambda_2 = \bigcap_{i=1}^{\infty}\{\gamma \mid \xi_i(\gamma) \geq H_2\}$$

has an uncertain measure $\mathcal{M}\{\Lambda_2\} = 1 - \epsilon$. Since $L_2(t)$ is an integrable and monotone decreasing function on $[0, +\infty)$, we have

$$\sum_{i=1}^{\infty} L_2(S_i(\gamma)) < \frac{1}{H_2}\int_0^{+\infty} L_2(s)ds < +\infty$$

for any $\gamma \in \Lambda_2$. Take

$$\delta = \exp\left(-(1 + H_1)\int_0^{+\infty} L_1(s)ds - \frac{1}{H_2}\int_0^{+\infty} L_2(s)ds\right)\varepsilon.$$

Then $|X_t(\gamma) - Y_t(\gamma)| \leq \varepsilon$ for any time t provided that $|X_0 - Y_0| \leq \delta$ and $\gamma \in \Lambda_1 \cap \Lambda_2$. It means

$$\mathcal{M}\left\{\sup_{t\geq 0}|X_t - Y_t| \leq \varepsilon\right\} \geq \mathcal{M}\{\Lambda_1 \cap \Lambda_2\} \geq 1 - \epsilon$$

as long as $|X_0 - Y_0| \leq \delta$. In other words,

$$\lim_{|X_0-Y_0|\to 0} \mathcal{M}\left\{\sup_{t\geq 0}|X_t - Y_t| \leq \varepsilon\right\} = 1,$$

and the uncertain differential equation with jumps (8.16) is stable in measure. The theorem is proved.

Almost Sure Stability

Definition 8.3 (*Ji and Ke [29]*) An uncertain differential equation with jumps

$$dX_t = f(t, X_t)dt + g(t, X_t)dC_t + h(t, X_t)dN_t \qquad (8.19)$$

is said to be almost surely stable if for any two solutions X_t and Y_t with different initial values X_0 and Y_0, we have

$$\mathcal{M} \left\{ \lim_{|X_0 - Y_0| \to 0} \sup_{t \geq 0} |X_t - Y_t| = 0 \right\} = 1. \tag{8.20}$$

Example 8.5 Consider an uncertain differential equation with jumps

$$dX_t = \mu dt + \sigma dC_t + \nu dN_t. \tag{8.21}$$

Since its solutions with different initial values X_0 and Y_0 are

$$X_t = X_0 + \mu t + \sigma C_t + \nu N_t,$$

$$Y_t = Y_0 + \mu t + \sigma C_t + \nu N_t,$$

respectively, we have

$$\sup_{t \geq 0} |X_t(\gamma) - Y_t(\gamma)| = |X_0 - Y_0|, \quad \forall \gamma \in \Gamma.$$

Then

$$\mathcal{M} \left\{ \lim_{|X_0 - Y_0| \to 0} \sup_{t \geq 0} |X_t - Y_t| = 0 \right\} = \mathcal{M} \left\{ \lim_{|X_0 - Y_0| \to 0} |X_0 - Y_0| = 0 \right\} = 1,$$

and the uncertain differential equation with jumps (8.21) is almost surely stable.

Example 8.6 Consider an uncertain differential equation with jumps

$$dX_t = \mu X_t dt + \sigma X_t dC_t + \nu X_t dN_t \tag{8.22}$$

where μ, σ, ν are positive real numbers. Since its solutions with different initial values X_0 and Y_0 are

$$X_t = X_0 \cdot \exp(\mu t + \sigma C_t) \cdot (1 + \nu)^{N_t},$$

$$Y_t = Y_0 \cdot \exp(\mu t + \sigma C_t) \cdot (1 + \nu)^{N_t},$$

respectively, we have

$$\sup_{t \geq 0} |X_t(\gamma) - Y_t(\gamma)| = |X_0 - Y_0| \cdot \exp(\mu t + \sigma C_t(\gamma)) \cdot (1 + \nu)^{N_t(\gamma)} = \infty$$

when $C_t(\gamma) > 0$ for all $t \in \mathfrak{R}$. Then

$$\mathcal{M}\left\{\lim_{|X_0 - Y_0| \to 0} \sup_{t \geq 0} |X_t - Y_t| = 0\right\} \leq 1 - \mathcal{M}\{C_t \geq 0, \forall t\} = 1/2,$$

and the uncertain differential equation with jumps (8.22) is not almost surely stable.

Theorem 8.5 (Ji and Ke [29]) *Assume the uncertain differential equation with jumps*

$$dX_t = f(t, X_t)dt + g(t, X_t)dC_t + h(t, X_t)dN_t \tag{8.23}$$

has a unique solution for each given initial value. Then it is almost surely stable if the coefficients $f(t, x)$ and $g(t, x)$ satisfy

$$|f(t, x) - f(t, y)| + |g(t, x) - g(t, y)| \leq L_1(t)|x - y|, \quad \forall x, y \in \mathfrak{R}, t \geq 0$$

for some integrable function $L_1(t)$ on $[0, +\infty)$, and the coefficient $h(t, x)$ satisfies

$$|h(t, x) - h(t, y)| \leq L_2(t)|x - y|, \quad \forall x, y \in \mathfrak{R}, t \geq 0$$

for some monotone and integrable function $L_2(t)$ on $[0, +\infty)$.

Proof Let X_t and Y_t be the solutions of the uncertain differential equation with jumps (8.23) with different initial values X_0 and Y_0, respectively. Then for a Lipschitz continuous sample path $C_t(\gamma)$, we have

$$X_t(\gamma) = X_0 + \int_0^t f(s, X_s(\gamma))ds + \int_0^t g(s, X_s(\gamma))dC_s(\gamma) + \int_0^t h(s, X_s(\gamma))dN_s(\gamma),$$

$$Y_t(\gamma) = Y_0 + \int_0^t f(s, Y_s(\gamma))ds + \int_0^t g(s, Y_s(\gamma))dC_s(\gamma) + \int_0^t h(s, Y_s(\gamma))dN_s(\gamma).$$

By the strong Lipschitz condition, we have

$$|X_t(\gamma) - Y_t(\gamma)|$$

$$\leq |X_0 - Y_0| + \int_0^t |f(s, X_s(\gamma)) - f(s, Y_s(\gamma))|ds$$

$$+ \int_0^t |g(s, X_s(\gamma)) - g(s, Y_s(\gamma))||dC_s(\gamma)| + \int_0^t |h(s, X_s(\gamma)) - h(s, Y_s(\gamma))|dN_s(\gamma)$$

$$\leq |X_0 - Y_0| + \int_0^t L_1(s)|X_s(\gamma) - Y_s(\gamma)|ds$$

$$+ \int_0^t L_1(s)K(\gamma)|X_s(\gamma) - Y_s(\gamma)|ds + \int_0^t L_2(s)|X_s(\gamma) - Y_s(\gamma)|dN_s(\gamma)$$

$$= |X_0 - Y_0| + \int_0^t (1 + K(\gamma)) L_1(s) |X_s(\gamma) - Y_s(\gamma)| ds$$

$$+ \int_0^t L_2(s) |X_s(\gamma) - Y_s(\gamma)| dN_s(\gamma)$$

where $K(\gamma)$ is the Lipschitz constant of $C_t(\gamma)$. Let $\xi_1, \xi_2, \ldots$ denote the interarrival times of the uncertain renewal process N_t. Write $S_0 = 0$ and $S_i = \xi_1 + \xi_2 + \ldots + \xi_i$ for $i \geq 1$. Then

$$|X_t(\gamma) - Y_t(\gamma)|$$

$$\leq |X_0 - Y_0| \cdot \exp\left((1 + K(\gamma)) \int_0^t L_1(s) ds \right) \cdot \prod_{i=1}^{N_t(\gamma)} (1 + L_2(S_i(\gamma)))$$

$$\leq |X_0 - Y_0| \cdot \exp\left((1 + K(\gamma)) \int_0^{+\infty} L_1(s) ds \right) \cdot \prod_{i=1}^{\infty} (1 + L_2(S_i(\gamma)))$$

$$\leq |X_0 - Y_0| \cdot \exp\left((1 + K(\gamma)) \int_0^{+\infty} L_1(s) ds + \sum_{i=1}^{\infty} L_2(S_i(\gamma)) \right)$$

for any $t \geq 0$. Thus

$$\sup_{t \geq 0} |X_t - Y_t| \leq |X_0 - Y_0| \cdot \exp\left((1 + K) \int_0^{+\infty} L_1(s) ds + \sum_{i=1}^{\infty} L_2(S_i) \right) \quad (8.24)$$

almost surely. We first consider

$$(1 + K) \int_0^{+\infty} L_1(s) ds.$$

Since the uncertain variable K is finite almost surely, and

$$\int_0^{+\infty} L_1(s) ds < +\infty,$$

we have

$$\mathcal{M}\left\{ (1 + K) \int_0^{+\infty} L_1(s) ds < +\infty \right\} = 1. \quad (8.25)$$

Now, we consider

$$\sum_{i=1}^{\infty} L_2(S_i).$$

Noting that

$$\sum_{i=1}^{\infty} L_2(S_i(\gamma)) \cdot \inf_{j \geq 1} \xi_j(\gamma) \leq \sum_{i=1}^{\infty} L_2(S_i(\gamma)) \cdot \xi_{i+1}(\gamma) \leq \int_0^{+\infty} L_2(s) \mathrm{d}s,$$

we have

$$\sum_{i=1}^{\infty} L_2(S_i(\gamma)) \leq \frac{1}{\inf_{j \geq 1} \xi_j(\gamma)} \int_0^{+\infty} L_2(s) \mathrm{d}s.$$

Since

$$\mathcal{M} \left\{ \gamma \in \Gamma \mid \inf_{j \geq 1} \xi_j(\gamma) > 0 \right\} = 1$$

and

$$\int_0^{+\infty} L_2(s) \mathrm{d}s < +\infty,$$

we have

$$\mathcal{M} \left\{ \sum_{i=1}^{\infty} L_2(S_i) < +\infty \right\} = 1. \tag{8.26}$$

By Eqs. (8.24)–(8.26), we have

$$\mathcal{M} \left\{ \lim_{|X_0 - Y_0| \to 0} \sup_{t \geq 0} |X_t - Y_t| = 0 \right\} = 1,$$

and the uncertain differential equation with jumps (8.23) is almost surely stable. The theorem is proved.

8.4   Yu's Stock Model

In 2012, Yu [88] supposed that the interest rate r_t is a constant and the stock price X_t follows an uncertain differential equation with jumps and proposed an uncertain stock model as follows:

$$\begin{cases} r_t = \mu_1 \\ \mathrm{d}X_t = \mu_2 X_t \mathrm{d}t + \sigma_2 X_t \mathrm{d}C_t + \nu_2 X_t \mathrm{d}N_t \end{cases} \tag{8.27}$$

where μ_1 is the riskless interest rate, μ_2, σ_2, and ν_2 are the log-drift, log-diffusion and jump size of the stock price, respectively, C_t is a canonical Liu process, and N_t is an uncertain renewal process. Note that the stock price is

$$X_t = X_0 \cdot \exp\left(\mu_2 t + \sigma_2 C_t\right) \left(1 + \nu_2\right)^{N_t} \qquad (8.28)$$

with an uncertainty distribution

$$\Upsilon_t(x) = \sup_{n \geq 0} \left(1 - \Psi\left(\frac{t}{n+1}\right)\right) \wedge \Phi\left(\frac{\ln(x) - \ln(X_0) - \mu_2 t - n\ln(1+\nu_2)}{\sigma_2 t}\right).$$

Here, Φ denotes the uncertainty distribution of standard normal uncertain variables, and Ψ denotes the uncertainty distribution of the interarrival times of the uncertain renewal process.

European Option Pricing Formulas

Consider a European call option of the stock model (8.27) with a strike price K and an expiration date T. Its price is determined by

$$f_c = \exp(-\mu_1 T) \cdot E[(X_T - K)^+].$$

Theorem 8.6 (Yu [88]) *The European call option price of the stock model (8.27) with a strike price K and an expiration date T is*

$$f_c = \exp(-\mu_1 T) \int_0^{+\infty} \Upsilon(x)dx \qquad (8.29)$$

where

$$\Upsilon(x) = \sup_{n \geq 0} \Psi\left(\frac{T}{n}\right) \wedge \left(1 - \Phi\left(\frac{\ln(K+x) - \ln(X_0) - \mu_2 T - n\ln(1+\nu_2)}{\sigma_2 T}\right)\right).$$

Here, Φ denotes the uncertainty distribution of standard normal uncertain variables, and Ψ denotes the uncertainty distribution of the interarrival times of the uncertain renewal process.

Proof Note that for each $x \geq 0$, we have

$$\mathcal{M}\left\{(X_T - K)^+ \geq x\right\}$$
$$= \mathcal{M}\left\{X_0 \exp(\mu_2 T + \sigma_2 C_T)(1 + \nu_2)^{N_T} \geq K + x\right\}$$
$$= \mathcal{M}\{\sigma_2 C_T + \ln(1+\nu_2)N_T \geq \ln(K+x) - \ln(X_0) - \mu_2 T\}$$
$$= \sup_{n \geq 0} \mathcal{M}\{N_T \geq n\} \wedge \mathcal{M}\left\{C_T \geq \frac{\ln(K+x) - \ln(X_0) - \mu_2 T - n\ln(1+\nu_2)}{\sigma_2}\right\}$$
$$= \sup_{n \geq 0} \Psi\left(\frac{T}{n}\right) \wedge \left(1 - \Phi\left(\frac{\ln(K+x) - \ln(X_0) - \mu_2 T - n\ln(1+\nu_2)}{\sigma_2 T}\right)\right)$$
$$= \Upsilon(x).$$

Then it follows from Theorem 2.7 that

$$f_c = \exp(-\mu_1 T) \int_0^{+\infty} \mathcal{M}\left\{(X_T - K)^+ \geq x\right\} dx$$

$$= \exp(-\mu_1 T) \int_0^{+\infty} \Upsilon(x) dx.$$

The theorem is proved.

Now consider a European put option of the stock model (8.27) with a strike price K and an expiration date T. Its price is determined by

$$f_p = \exp(-\mu_1 T) \cdot E[(K - X_T)^+].$$

Theorem 8.7 (Yu [88]) *The European put option price of the stock model (8.27) with a strike price K and an expiration date T is*

$$f_p = \exp(-\mu_1 T) \int_0^K \Upsilon(x) dx \tag{8.30}$$

where

$$\Upsilon(x) = \sup_{n \geq 0}\left(1 - \Psi\left(\frac{T}{n+1}\right)\right) \wedge \Phi\left(\frac{\ln(K - x) - \ln(X_0) - \mu_2 T - n \ln(1 + \nu_2)}{\sigma_2 T}\right).$$

Here, Φ denotes the uncertainty distribution of standard normal uncertain variables, and Ψ denotes the uncertainty distribution of the interarrival times of the uncertain renewal process.

Proof Note that for each $x \in [0, K]$, we have

$$\mathcal{M}\left\{(K - X_T)^+ \geq x\right\}$$
$$= \mathcal{M}\left\{X_0 \exp(\mu_2 T + \sigma_2 C_T)(1 + \nu_2)^{N_T} \leq K - x\right\}$$
$$= \mathcal{M}\{\sigma_2 C_T + \ln(1 + \nu_2)N_T \leq \ln(K - x) - \ln(X_0) - \mu_2 T\}$$
$$= \sup_{n \geq 0} \mathcal{M}\{N_T \leq n\} \wedge \mathcal{M}\left\{C_T \leq \frac{\ln(K - x) - \ln(X_0) - \mu_2 T - n \ln(1 + \nu_2)}{\sigma_2}\right\}$$
$$= \sup_{n \geq 0}\left(1 - \Psi\left(\frac{T}{n+1}\right)\right) \wedge \Phi\left(\frac{\ln(K - x) - \ln(X_0) - \mu_2 T - n \ln(1 + \nu_2)}{\sigma_2 T}\right)$$
$$= \Upsilon(x).$$

Then it follows from Theorem 2.7 that

$$
\begin{aligned}
f_c &= \exp(-\mu_1 T) \int_0^{+\infty} \mathcal{M}\left\{(K - X_T)^+ \geq x\right\} \mathrm{d}x \\
&= \exp(-\mu_1 T) \int_0^K \mathcal{M}\left\{(K - X_T)^+ \geq x\right\} \mathrm{d}x \\
&= \exp(-\mu_1 T) \int_0^K \Upsilon(x) \mathrm{d}x.
\end{aligned}
$$

The theorem is proved.

Chapter 9
Multi-Dimensional Uncertain Differential Equation

Multi-dimensional uncertain differential equation is a system of uncertain differential equations. The emphases of this chapter are on multi-dimensional Liu process, multi-dimensional uncertain calculus, and multi-dimensional uncertain differential equation. For simplicity, we employ the infinite norm in this chapter and write

$$|x| = \bigvee_{i=1}^{n} |x_i|, \quad |A| = \bigvee_{i=1}^{m} \sum_{j=1}^{n} |a_{ij}|$$

for an n-dimensional vector $x = (x_1, x_2, \ldots, x_n)^T$ and an $m \times n$ matrix $A = [a_{ij}]$, respectively.

9.1 Multi-Dimensional Canonical Liu Process

Definition 9.1 (*Zhang and Chen [90]*) Let $C_{it}, i = 1, 2, \ldots, n$ be independent canonical Liu processes. Then $C_t = (C_{1t}, C_{2t}, \ldots, C_{nt})^T$ is called an n-dimensional canonical Liu process.

Zhang and Chen [90] showed that C_t is a stationary independent increment multi-dimensional uncertain process.

Theorem 9.1 (Zhang and Chen [90]) *Almost all sample paths of a multi-dimensional canonical Liu process are Lipschitz continuous.*

Proof Without loss of generality, we assume $C_t = (C_{1t}, C_{2t}, \ldots, C_{nt})^T$ is an n-dimensional canonical Liu process. Write

$$\Lambda_i = \{\gamma \in \Gamma \mid C_{it}(\gamma) \text{ is Lipschitz continuous}\}$$

for $i = 1, 2, \ldots, n$. Since almost all the sample paths of C_{it} are Lipschitz continuous, we have $\mathcal{M}\{\Lambda_i\} = 1$. Write

© Springer-Verlag Berlin Heidelberg 2016

K. Yao, *Uncertain Differential Equations*,
Springer Uncertainty Research, DOI 10.1007/978-3-662-52729-0_9

$$\Lambda = \bigcap_{i=1}^{n} \Lambda_i,$$

then

$$\mathcal{M}\{\Lambda\} = 1 - \mathcal{M}\left\{\bigcup_{i=1}^{n} \Lambda_i^c\right\} \geq 1 - \sum_{i=1}^{n} \mathcal{M}\{\Lambda_i^c\} = 1.$$

For each $\gamma \in \Lambda$, we have

$$|C_{t_1}(\gamma) - C_{t_2}(\gamma)| = \bigvee_{i=1}^{n} |C_{it_1}(\gamma) - C_{it_2}(\gamma)| \leq \bigvee_{i=1}^{n} K_i(\gamma)|t_1 - t_2|,$$

where $K_i(\gamma)$ are the Lipschitz constants of the sample paths $C_{it}(\gamma)$, $i = 1, 2, \ldots, n$. Thus almost all sample paths of C_t are Lipschitz continuous. The theorem is proved.

Theorem 9.2 (Su et al. [62]) *Let C_t be an n-dimensional canonical Liu process on an uncertainty space $(\Gamma, \mathcal{L}, \mathcal{M})$. Then there exists an uncertain variable K such that $K(\gamma)$ is a Lipschitz constant of the sample path $C_t(\gamma)$ for each $\gamma \in \Gamma$, and*

$$\lim_{x \to +\infty} \mathcal{M}\{\gamma \in \Gamma \mid K(\gamma) \leq x\} = 1. \tag{9.1}$$

Proof For each $\gamma \in \Gamma$, set

$$K(\gamma) = \bigvee_{i=1}^{n} K_i(\gamma)$$

where $K_i(\gamma)$ are the Lipschitz constants of the sample paths $C_{it}(\gamma)$, $i = 1, 2, \ldots$, n. Then K is an uncertain variable, and $K(\gamma)$ is a Lipschitz constant of the sample path $C_t(\gamma)$ according to Theorem 9.1. Noting that

$$\lim_{x \to +\infty} \mathcal{M}\{\gamma \in \Gamma \mid K_i(\gamma) > x\} = 0, \quad i = 1, 2, \ldots, n$$

according to Theorem 5.3 and

$$\mathcal{M}\{\gamma \in \Gamma \mid K(\gamma) \leq x\} \geq 1 - \sum_{i=1}^{n} \mathcal{M}\{\gamma \in \Gamma \mid K_i(\gamma) > x\}$$

according to the duality and subadditivity of uncertain measure, we have

$$\lim_{x \to +\infty} \mathcal{M}\{\gamma \in \Gamma \mid K(\gamma) \leq x\} = 1.$$

The theorem is proved.

9.2 Multi-Dimensional Liu Integral

Definition 9.2 (*Yao [79]*) An $m \times n$ matrix $X_t = [X_{ijt}]$ is called an uncertain matrix process if its elements X_{ijt} are uncertain processes for $i = 1, 2, \ldots, m, j = 1, 2, \ldots, n$.

Definition 9.3 (*Yao [79]*) Let $C_t = (C_{1t}, C_{2t}, \ldots, C_{nt})^T$ be an n-dimensional canonical Liu process, and $X_t = [X_{ijt}]$ be an $m \times n$ uncertain matrix process whose elements X_{ijt} are integrable uncertain processes. Then the Liu integral of X_t with respect to C_t on the interval $[a, b]$ is defined by

$$\int_a^b X_t dC_t = \begin{pmatrix} \sum_{j=1}^{n} \int_a^b X_{1jt} dC_{jt} \\ \sum_{j=1}^{n} \int_a^b X_{2jt} dC_{jt} \\ \vdots \\ \sum_{j=1}^{n} \int_a^b X_{mjt} dC_{jt} \end{pmatrix}. \tag{9.2}$$

In this case, X_t is said to be Liu integrable with respect to C_t.

Example 9.1 Let $C_t = (C_{1t}, C_{2t}, \ldots, C_{nt})^T$ be an n-dimensional canonical Liu process, and $X_t = (X_{1t}, X_{2t}, \ldots, X_{nt})$ be an n-dimensional Liu integrable uncertain process. Then

$$\int_a^b X_t dC_t = \sum_{j=1}^{n} \int_a^b X_{jt} dC_{jt}.$$

This type of uncertain calculus, named multifactor uncertain calculus, was first proposed by Liu and Yao [42].

Example 9.2 Let C_t be a canonical Liu process, and $X_t = (X_{1t}, X_{2t}, \ldots, X_{nt})^T$ be an n-dimensional Liu integrable uncertain process. Then

$$\int_a^b X_t dC_t = \begin{pmatrix} \int_a^b X_{1t} dC_t \\ \int_a^b X_{2t} dC_t \\ \vdots \\ \int_a^b X_{mt} dC_t \end{pmatrix}.$$

Example 9.3 Let $C_t = (C_{1t}, C_{2t})^T$ be a 2-dimensional canonical Liu process, and

$$X_t = \begin{pmatrix} C_{1t} & 0 \\ 0 & C_{2t} \end{pmatrix}$$

be an uncertain matrix process. Then

$$\int_0^t X_s dC_s = \frac{1}{2} \begin{pmatrix} C_{1t}^2 \\ C_{2t}^2 \end{pmatrix}.$$

Theorem 9.3 (Ji and Zhou [31]) *Let C_t be an n-dimensional canonical Liu process, and X_t be an $m \times n$ Liu integrable uncertain matrix process. Then*

$$\left| \int_a^b X_t(\gamma) dC_t(\gamma) \right| \le K(\gamma) \int_a^b |X_t(\gamma)| dt, \quad \forall \gamma \in \Gamma \tag{9.3}$$

where $K(\gamma)$ is the Lipschitz constant of the sample path $C_t(\gamma)$.

Proof For each $\gamma \in \Gamma$, by using the infinite norm, we have

$$\left| \int_a^b X_t(\gamma) dC_t(\gamma) \right| = \bigvee_{i=1}^m \left| \sum_{j=1}^n \int_a^b X_{ijt}(\gamma) dC_{jt}(\gamma) \right|$$

$$\le \bigvee_{i=1}^m \sum_{j=1}^n \left| \int_a^b X_{ijt}(\gamma) dC_{jt}(\gamma) \right|.$$

Since

$$\left| \int_a^b X_{ijt}(\gamma) dC_{jt}(\gamma) \right| \le K(\gamma) \int_a^b |X_{ijt}(\gamma)| dt$$

according to Theorem 5.5, we get

$$\left| \int_a^b X_t(\gamma) dC_t(\gamma) \right| \le \bigvee_{i=1}^m \sum_{j=1}^n K(\gamma) \int_a^b |X_{ijt}(\gamma)| dt$$

$$= K(\gamma) \bigvee_{i=1}^m \int_a^b \sum_{j=1}^n |X_{ijt}(\gamma)| dt \le K(\gamma) \int_a^b \bigvee_{i=1}^m \sum_{j=1}^n |X_{ijt}(\gamma)| dt.$$

Noting that

$$|X_t(\gamma)| = \bigvee_{i=1}^m \sum_{j=1}^n |X_{ijt}(\gamma)|,$$

we immediately have

$$\left| \int_a^b X_t(\gamma) dC_t(\gamma) \right| \le K(\gamma) \int_a^b |X_t(\gamma)| dt.$$

The theorem is proved.

Theorem 9.4 (Yao [79]) *Let C_t be an n-dimensional canonical Liu process, and X_t be an $m \times n$ Liu integrable uncertain matrix process on $[a, b]$. Then X_t is Liu integrable with respect to C_t on each subinterval of $[a, b]$. Moreover, if $c \in [a, b]$, then*

$$\int_a^b X_t dC_t = \int_a^c X_t dC_t + \int_c^b X_t dC_t. \tag{9.4}$$

Proof Since X_t is Liu integrable with respect to C_t on $[a, b]$, the uncertain process X_{ijt} is Liu integrable with respect to C_{jt} on $[a, b]$. Then X_{ijt} is Liu integrable with respect to C_{jt} on each subinterval of $[a, b]$ according to Theorem 5.6. By Definition 9.3, the uncertain matrix process X_t is Liu integrable with respect to C_t on each subinterval of $[a, b]$. Next, for a point $c \in [a, b]$, we have

$$\int_a^b X_t dC_t$$

$$= \begin{pmatrix} \sum_{j=1}^n \int_a^b X_{1jt} dC_{jt} \\ \sum_{j=1}^n \int_a^b X_{2jt} dC_{jt} \\ \vdots \\ \sum_{j=1}^n \int_a^b X_{mjt} dC_{jt} \end{pmatrix} = \begin{pmatrix} \sum_{j=1}^n \int_a^c X_{1jt} dC_{jt} + \sum_{j=1}^n \int_c^b X_{1jt} dC_{jt} \\ \sum_{j=1}^n \int_a^c X_{2jt} dC_{jt} + \sum_{j=1}^n \int_c^b X_{2jt} dC_{jt} \\ \vdots \\ \sum_{j=1}^n \int_a^c X_{mjt} dC_{jt} + \sum_{j=1}^n \int_c^b X_{mjt} dC_{jt} \end{pmatrix}$$

$$= \begin{pmatrix} \sum_{j=1}^n \int_a^c X_{1jt} dC_{jt} \\ \sum_{j=1}^n \int_a^c X_{2jt} dC_{jt} \\ \vdots \\ \sum_{j=1}^n \int_a^c X_{mjt} dC_{jt} \end{pmatrix} + \begin{pmatrix} \sum_{j=1}^n \int_c^b X_{1jt} dC_{jt} \\ \sum_{j=1}^n \int_c^b X_{2jt} dC_{jt} \\ \vdots \\ \sum_{j=1}^n \int_c^b X_{mjt} dC_{jt} \end{pmatrix}$$

$$= \int_a^c X_t dC_t + \int_c^b X_t dC_t.$$

The theorem is proved.

Theorem 9.5 (Yao [79], Linearity of Multi-dimensional Liu Integral) *Let C_t be an n-dimensional canonical Liu process, and X_t and Y_t be two $m \times n$ Liu integrable uncertain matrix processes on $[a, b]$. Then*

$$\int_a^b (\alpha X_t + \beta Y_t) dC_t = \alpha \int_a^b X_t dC_t + \beta \int_a^b Y_t dC_t \qquad (9.5)$$

for any real numbers α and β.

Proof It follows from Definition 9.3 that

$$\int_a^b (\alpha X_t + \beta Y_t) dC_t$$

$$= \begin{pmatrix} \sum_{j=1}^n \int_a^b (\alpha X_{1jt} + \beta Y_{1jt}) dC_{jt} \\ \sum_{j=1}^n \int_a^b (\alpha X_{2jt} + \beta Y_{2jt}) dC_{jt} \\ \vdots \\ \sum_{j=1}^n \int_a^b (\alpha X_{mjt} + \beta Y_{mjt}) dC_{jt} \end{pmatrix}$$

$$= \begin{pmatrix} \alpha \sum_{j=1}^n \int_a^b X_{1jt} dC_{jt} + \beta \sum_{j=1}^n \int_a^b Y_{1jt} dC_{jt} \\ \alpha \sum_{j=1}^n \int_a^b X_{2jt} dC_{jt} + \beta \sum_{j=1}^n \int_a^b Y_{2jt} dC_{jt} \\ \vdots \\ \alpha \sum_{j=1}^n \int_a^b X_{mjt} dC_{jt} + \beta \sum_{j=1}^n \int_a^b Y_{mjt} dC_{jt} \end{pmatrix}$$

$$= \alpha \begin{pmatrix} \sum_{j=1}^n \int_a^b X_{1jt} dC_{jt} \\ \sum_{j=1}^n \int_a^b X_{2jt} dC_{jt} \\ \vdots \\ \sum_{j=1}^n \int_a^b X_{mjt} dC_{jt} \end{pmatrix} + \beta \begin{pmatrix} \sum_{j=1}^n \int_a^b Y_{1jt} dC_{jt} \\ \sum_{j=1}^n \int_a^b Y_{2jt} dC_{jt} \\ \vdots \\ \sum_{j=1}^n \int_a^b Y_{mjt} dC_{jt} \end{pmatrix}$$

$$= \alpha \int_a^b X_t dC_t + \beta \int_a^b Y_t dC_t.$$

The theorem is proved.

9.3 Multi-Dimensional Liu Process

Definition 9.4 (*Yao [79]*) Let C_t be an n-dimensional canonical Liu process, μ_t be an m-dimensional time integrable uncertain process, and σ_t be an $m \times n$ Liu integrable uncertain matrix process. Then the m-dimensional uncertain process

$$Z_t = Z_0 + \int_0^t \mu_s ds + \int_0^t \sigma_s dC_s \tag{9.6}$$

is called an m-dimensional Liu process, and it has a Liu differential

$$dZ_t = \mu_t dt + \sigma_t dC_t. \tag{9.7}$$

Example 9.4 Let $C_t = (C_{1t}, C_{2t}, \ldots, C_{nt})^T$ be an n-dimensional canonical Liu process. Since

$$C_t = \int_0^t dC_s,$$

it is an n-dimensional Liu process and has a Liu differential dC_t.

Example 9.5 Let $C_t = (C_{1t}, C_{2t}, \ldots, C_{nt})^T$ be an n-dimensional canonical Liu process. Since the n-dimensional uncertain process tC_t satisfies

$$tC_t = \int_0^t C_s ds + \int_0^t s dC_s,$$

it is an n-dimensional Liu process and has a Liu differential

$$d(tC_t) = C_t dt + t dC_t.$$

Example 9.6 Let $C_t = (C_{1t}, C_{2t})^T$ be a 2-dimensional canonical Liu process. Since the 2-dimensional uncertain process $Z_t = (C_{1t}^2, C_{2t}^2)^T$ satisfies

$$\begin{pmatrix} C_{1t}^2 \\ C_{2t}^2 \end{pmatrix} = 2 \int_0^t \begin{pmatrix} C_{1s} & 0 \\ 0 & C_{2s} \end{pmatrix} d \begin{pmatrix} C_{1s} \\ C_{2s} \end{pmatrix},$$

it is a 2-dimensional Liu process and has a Liu differential

$$dZ_t = 2 \begin{pmatrix} C_{1t} & 0 \\ 0 & C_{2t} \end{pmatrix} dC_t.$$

Fundamental Theorem

Theorem 9.6 (Yao [79], Fundamental Theorem) *Let* $C_t = (C_{1t}, C_{2t}, \ldots, C_{nt})^T$ *be an n-dimensional canonical Liu process, and* $h(t, c) = (h_1(t, c), h_2(t, c), \ldots, h_m(t, c))^T$ *be an m-dimensional continuously differentiable function, where*

$c = (c_1, c_2, \ldots, c_n)^T$ *is an n-dimensional vector. Then the m-dimensional uncertain process* $Z_t = h(t, C_t)$ *is an m-dimensional Liu process, and it has a Liu differential*

$$\mathrm{d}Z_t = \frac{\partial h}{\partial t}(t, C_t)\mathrm{d}t + \frac{\partial h}{\partial c}(t, C_t)\mathrm{d}C_t \tag{9.8}$$

where

$$\frac{\partial h}{\partial t}(t, C_t) = \begin{pmatrix} \dfrac{\partial h_1}{\partial t}(t, C_{1t}, \ldots, C_{nt}) \\[2mm] \dfrac{\partial h_2}{\partial t}(t, C_{1t}, \ldots, C_{nt}) \\[2mm] \vdots \\[2mm] \dfrac{\partial h_m}{\partial t}(t, C_{1t}, \ldots, C_{nt}) \end{pmatrix} \tag{9.9}$$

and

$$\frac{\partial h}{\partial c}(t, C_t) = \begin{pmatrix} \dfrac{\partial h_1}{\partial c_1}(t, C_{1t}, \ldots, C_{nt}) & \cdots & \dfrac{\partial h_1}{\partial c_n}(t, C_{1t}, \ldots, C_{nt}) \\[2mm] \dfrac{\partial h_2}{\partial c_1}(t, C_{1t}, \ldots, C_{nt}) & \cdots & \dfrac{\partial h_2}{\partial c_n}(t, C_{1t}, \ldots, C_{nt}) \\[2mm] \vdots & \ddots & \vdots \\[2mm] \dfrac{\partial h_m}{\partial c_1}(t, C_{1t}, \ldots, C_{nt}) & \cdots & \dfrac{\partial h_m}{\partial c_n}(t, C_{1t}, \ldots, C_{nt}) \end{pmatrix}. \tag{9.10}$$

Proof Note that $\Delta C_t = C_{t+\Delta t} - C_t$ is an infinitesimal with the same order as Δt. By using Taylor series expansion, we get a first-order approximation

$$\Delta Z_t = \frac{\partial h}{\partial t}(t, C_t)\Delta t + \frac{\partial h}{\partial c}(t, C_t)\Delta C_t.$$

Letting $\Delta t \to 0$, we have

$$\mathrm{d}Z_t = \frac{\partial h}{\partial t}(t, C_t)\mathrm{d}t + \frac{\partial h}{\partial c}(t, C_t)\mathrm{d}C_t.$$

The theorem is proved.

Example 9.7 Consider the Liu differential of the multi-dimensional Liu process $Z_t = \mu t + \sigma C_t$. In this case, we assume $h(t, c) = \mu t + \sigma c$. Since

$$\frac{\partial h}{\partial t}(t, c) = \mu, \quad \frac{\partial h}{\partial c}(t, c) = \sigma,$$

we have

$$\mathrm{d}Z_t = \mu \mathrm{d}t + \sigma \mathrm{d}C_t.$$

Example 9.8 Consider the Liu differential of the multi-dimensional Liu process $Z_t = tC_t$. In this case, we assume $h(t, c) = tc$. Since

$$\frac{\partial h}{\partial t}(t, c) = c, \quad \frac{\partial h}{\partial c}(t, c) = tI$$

where I is the identity matrix, we have

$$dZ_t = C_t dt + t dC_t.$$

Example 9.9 Consider the Liu differential of the 2-dimensional Liu process $Z_t = (C_{1t}^2, C_{2t}^2)^T$. In this case, we assume $h(t, c) = (c_1^2, c_2^2)^T$. Since

$$\frac{\partial h}{\partial t}(t, c) = \begin{pmatrix} 0 \\ 0 \end{pmatrix}, \quad \frac{\partial h}{\partial c}(t, c) = 2\begin{pmatrix} c_1 & 0 \\ 0 & c_2 \end{pmatrix},$$

we have

$$dZ_t = 2\begin{pmatrix} C_{1t} & 0 \\ 0 & C_{2t} \end{pmatrix} dC_t.$$

Theorem 9.7 (Liu and Yao [42]) *Let* $(C_{1t}, C_{2t}, \ldots, C_{nt})^T$ *be an n-dimensional canonical Liu process, and* $h(t, c_1, c_2, \ldots, c_n)$ *be a continuously differentiable function. Then the uncertain process* $Z_t = h(t, C_{1t}, C_{2t}, \ldots, C_{nt})$ *has a Liu differential*

$$dZ_t = \frac{\partial h}{\partial t}(t, C_{1t}, C_{2t}, \ldots, C_{nt})dt + \sum_{i=1}^{n} \frac{\partial h}{\partial c_i}(t, C_{1t}, C_{2t}, \ldots, C_{nt})dC_{it}. \quad (9.11)$$

Proof This theorem is just a special case of Theorem 9.6 that $m = 1$.

Theorem 9.8 *Let* X_t *be a multi-dimensional Liu process such that* $dX_t = \mu_t dt + \sigma_t dC_t$, *and* $h(t, x)$ *be a multi-dimensional continuously differentiable function. Then* $Z_t = h(t, X_t)$ *is also a multi-dimensional Liu process, and it has a Liu differential*

$$dZ_t = \left(\frac{\partial h}{\partial t}(t, X_t) + \frac{\partial h}{\partial x}(t, X_t) \cdot \mu_t\right) dt + \frac{\partial h}{\partial x}(t, X_t) \cdot \sigma_t dC_t. \quad (9.12)$$

Proof Note that

$$\Delta X_t = X_{t+\Delta t} - X_t = \mu_t \Delta t + \sigma_t \Delta C_t$$

is an infinitesimal with the same order as Δt. By using Taylor series expansion, we get a first-order approximation

$$\Delta Z_t = \frac{\partial h}{\partial t}(t, X_t)\Delta t + \frac{\partial h}{\partial x}(t, X_t)\Delta X_t$$
$$= \left(\frac{\partial h}{\partial t}(t, X_t) + \frac{\partial h}{\partial x}(t, X_t) \cdot \mu_t\right)\Delta t + \frac{\partial h}{\partial x}(t, X_t) \cdot \sigma_t \Delta C_t.$$

Letting $\Delta t \to 0$, we have

$$dZ_t = \left(\frac{\partial h}{\partial t}(t, X_t) + \frac{\partial h}{\partial x}(t, X_t) \cdot \mu_t\right)dt + \frac{\partial h}{\partial x}(t, X_t) \cdot \sigma_t dC_t.$$

The theorem is proved.

Theorem 9.9 (Yao [79]) *Let $X_{1t}, X_{2t}, \ldots, X_{nt}$ be multi-dimensional Liu processes, and $h(t, x_1, x_2, \ldots, x_n)$ be a multi-dimensional continuously differentiable function. Then the multi-dimensional uncertain process $X_t = h(t, X_{1t}, X_{2t}, \ldots, X_{nt})$ has an uncertain differential*

$$dX_t = \frac{\partial h}{\partial t}(t, X_{1t}, \ldots, X_{nt})dt + \sum_{i=1}^{n} \frac{\partial h}{\partial x_i}(t, X_{1t}, \ldots, X_{nt})dX_{it}. \qquad (9.13)$$

Proof Since the function h is continuously differentiable, by using Taylor series expansion, we get a first-order approximation

$$\Delta X_t = \frac{\partial h}{\partial t}(t, X_{1t}, \ldots, X_{nt})\Delta t + \sum_{i=1}^{n} \frac{\partial h}{\partial x_i}(t, X_{1t}, \ldots, X_{nt})\Delta X_{it}.$$

Letting $\Delta t \to 0$, we have

$$dX_t = \frac{\partial h}{\partial t}(t, X_{1t}, \ldots, X_{nt})dt + \sum_{i=1}^{n} \frac{\partial h}{\partial x_i}(t, X_{1t}, \ldots, X_{nt})dX_{it}.$$

The theorem is proved.

Integration by Parts

Theorem 9.10 (Yao [79], Integration by Parts) *Let X_t and Y_t be two m-dimensional Liu processes. Then*
$$d(X_t^T Y_t) = Y_t^T dX_t + X_t^T dY_t. \qquad (9.14)$$

Proof Taking $h(t, x, y) = x^T y$ in Theorem 9.9, we have

$$\frac{\partial h}{\partial t} = 0, \quad \frac{\partial h}{\partial x} = y^T, \quad \frac{\partial h}{\partial y} = x^T.$$

Then
$$d(X_t^T Y_t) = Y_t^T dX_t + X_t^T dY_t,$$

and the theorem is proved.

9.4 Multi-Dimensional Uncertain Differential Equation

Definition 9.5 (*Yao [79]*) Suppose that C_t is an n-dimensional canonical Liu process, $f(t, x)$ is a vector-valued function from $T \times \mathfrak{R}^n$ to $\mathfrak{R}^m$, and $g(t, x)$ is a matrix-valued function from $T \times \mathfrak{R}^n$ to the set of $m \times n$ matrices. Then

$$dX_t = f(t, X_t)dt + g(t, X_t)dC_t \qquad (9.15)$$

is called an m-dimensional uncertain differential equation driven by an n-dimensional canonical Liu process. An m-dimensional uncertain process that satisfies (9.15) identically at each time t is called a solution of the multi-dimensional uncertain differential equation.

Remark 9.1 The multi-dimensional uncertain differential equation (9.15) is equivalent to the multi-dimensional uncertain integral equation

$$X_s = X_0 + \int_0^s f(t, X_t)dt + \int_0^s g(t, X_t)dC_t.$$

Apparently, the solution of a multi-dimensional uncertain differential equation is a multi-dimensional Liu process.

Remark 9.2 When $m = 1$, the uncertain differential equation (9.15) degenerates to the multifactor uncertain differential equation

$$dX_t = f(t, X_t)dt + \sum_{i=1}^n g_i(t, X_t)dC_{it}$$

which was studied by Li et al. [34]. For example, the uncertain differential equation

$$dX_t = X_t dt + (X_t, X_t)dC_t = X_t dt + X_t dC_{1t} + X_t dC_{2t}, \quad X_0 = 1$$

has a solution

$$X_t = \exp(t + C_{1t} + C_{2t}).$$

Theorem 9.11 (Ji and Zhou [31]) *Let C_t be an n-dimensional canonical Liu process, U_t be an m-dimensional time integrable function, and V_t be an $m \times n$ Liu integrable matrix function. Then the m-dimensional uncertain differential equation*

$$dX_t = U_t dt + V_t dC_t \tag{9.16}$$

has a solution

$$X_t = X_0 + \int_0^t U_s ds + \int_0^t V_s dC_s. \tag{9.17}$$

Proof Taking differentiation operations on both sides of Eq. (9.17), we have

$$dX_t = U_t dt + V_t dC_t.$$

The theorem is proved.

Example 9.10 Let $C_t = (C_{1t}, C_{2t})^T$ be a 2-dimensional canonical Liu process. Consider a 2-dimensional uncertain differential equation

$$dX_t = \begin{pmatrix} 0 \\ 1 \end{pmatrix} dt + \begin{pmatrix} 1 & 1 \\ -1 & 1 \end{pmatrix} dC_t$$

with an initial value $X_0 = (1, 0)^T$. According to Theorem 9.11, it has a solution

$$X_t = \begin{pmatrix} X_{1t} \\ X_{2t} \end{pmatrix} = \begin{pmatrix} 1 + C_{1t} + C_{2t} \\ t - C_{1t} + C_{2t} \end{pmatrix}.$$

Theorem 9.12 (Ji and Zhou [31]) *Let C_t be a canonical Liu process, and U and V be two $m \times m$ matrices satisfying $UV = VU$. Then the m-dimensional uncertain differential equation*

$$dX_t = U X_t dt + V X_t dC_t \tag{9.18}$$

has a solution

$$X_t = \exp(tU + C_t V) \cdot X_0. \tag{9.19}$$

Proof The m-dimensional uncertain process X_t can also be written as follows:

$$X_t = \sum_{n=0}^{\infty} \frac{1}{n!} (tU + C_t V)^n \cdot X_0.$$

Since $UV = VU$, taking differentiation operations on both sides of Eq. (9.19), we have

$$dX_t = U \sum_{n=1}^{\infty} \frac{1}{(n-1)!} (tU + C_t V)^{n-1} \cdot X_0 dt$$

$$+ V \sum_{n=1}^{\infty} \frac{1}{(n-1)!} (tU + C_t V)^{n-1} \cdot X_0 dC_t$$

$$= U \sum_{n=0}^{\infty} \frac{1}{n!} (tU + C_t V)^n \cdot X_0 dt + V \sum_{n=0}^{\infty} \frac{1}{n!} (tU + C_t V)^n \cdot X_0 dC_t$$
$$= U \exp(tU + C_t V) \cdot X_0 dt + V \exp(tU + C_t V) \cdot X_0 dC_t$$
$$= U X_t dt + V X_t dC_t.$$

The theorem is proved.

Example 9.11 Let C_t be a canonical Liu process. Consider a 2-dimensional uncertain differential equation

$$dX_t = \begin{pmatrix} 0 & 1 \\ -1 & 0 \end{pmatrix} X_t dC_t$$

with an initial value $X_0 = (1, 0)^T$. In this case, we have $U = 0$ and

$$V = \begin{pmatrix} 0 & 1 \\ -1 & 0 \end{pmatrix}.$$

Since $UV = VU = 0$, the 2-dimensional uncertain differential equation has a solution

$$X_t = \exp(C_t V) \cdot X_0 = \exp\begin{pmatrix} 0 & C_t \\ -C_t & 0 \end{pmatrix} \cdot \begin{pmatrix} 1 \\ 0 \end{pmatrix}$$
$$= \begin{pmatrix} \cos(C_t) & \sin(C_t) \\ -\sin(C_t) & \cos(C_t) \end{pmatrix} \cdot \begin{pmatrix} 1 \\ 0 \end{pmatrix} = \begin{pmatrix} \cos(C_t) \\ -\sin(C_t) \end{pmatrix}.$$

Example 9.12 Let C_t be a canonical Liu process. Consider a 2-dimensional uncertain differential equation

$$dX_t = \begin{pmatrix} 1 & 0 \\ 0 & 1 \end{pmatrix} X_t dt + \begin{pmatrix} 1 & 1 \\ -1 & 1 \end{pmatrix} X_t dC_t$$

with an initial value $X_0 = (0, 1)^T$. In this case, we have

$$U = \begin{pmatrix} 1 & 0 \\ 0 & 1 \end{pmatrix}, \qquad V = \begin{pmatrix} 1 & 1 \\ -1 & 1 \end{pmatrix}.$$

Since $UV = VU = V$, the 2-dimensional uncertain differential equation has a solution

$$X_t = \exp(tU + C_t V) \cdot X_0 = \exp\begin{pmatrix} t & 0 \\ 0 & t \end{pmatrix} \cdot \exp\begin{pmatrix} C_t & C_t \\ -C_t & C_t \end{pmatrix} \cdot \begin{pmatrix} 0 \\ 1 \end{pmatrix}$$

$$= \begin{pmatrix} \exp(t) & 0 \\ 0 & \exp(t) \end{pmatrix} \cdot \begin{pmatrix} \exp(C_t)\cos(C_t) & \exp(C_t)\sin(C_t) \\ -\exp(C_t)\sin(C_t) & \exp(C_t)\cos(C_t) \end{pmatrix} \cdot \begin{pmatrix} 0 \\ 1 \end{pmatrix}$$

$$= \begin{pmatrix} \exp(t + C_t)\sin(C_t) \\ \exp(t + C_t)\cos(C_t) \end{pmatrix}.$$

Existence and Uniqueness Theorem

Theorem 9.13 (Ji and Zhou [31]) *The multi-dimensional uncertain differential equation*

$$dX_t = f(t, X_t)dt + g(t, X_t)dC_t \tag{9.20}$$

with an initial value X_0 has a unique solution if the coefficients $f(t, x)$ and $g(t, x)$ satisfy the linear growth condition

$$|f(t, x)| + |g(t, x)| \le L(1 + |x|), \quad \forall x \in \Re^m, t \ge 0 \tag{9.21}$$

and the Lipschitz condition

$$|f(t, x) - f(t, y)| + |g(t, x) - g(t, y)| \le L|x - y|, \quad \forall x, y \in \Re^m, t \ge 0 \tag{9.22}$$

for some constant L.

Proof The proof of this theorem is similar to that of Theorem 6.7, so here it is introduced briefly. For any given real number T, consider the solution on the interval $[0, T]$. For each $\gamma \in \Gamma$, define $X_t^{(0)} = X_0$,

$$X_t^{(n+1)}(\gamma) = X_0 + \int_0^t f\left(s, X_s^{(n)}(\gamma)\right) ds + \int_0^t g\left(s, X_s^{(n)}(\gamma)\right) dC_s(\gamma)$$

and

$$Q_t^{(n)}(\gamma) = \sup_{0 \le s \le t} \left| X_s^{(n+1)}(\gamma) - X_s^{(n)}(\gamma) \right|$$

for $n = 1, 2, \ldots$ By the induction method, we have

$$Q_t^{(n)}(\gamma) \le (1 + |X_0|) \frac{L^{n+1}(1 + K(\gamma))^{n+1}}{(n+1)!} t^{n+1} \tag{9.23}$$

for almost every $\gamma \in \Gamma$ and for every nonnegative integer n, where $K(\gamma)$ is the Lipschitz constant of $C_t(\gamma)$. Since the right term of (9.23) satisfies

$$\sum_{n=0}^{\infty} (1 + |X_0|) \frac{L^{n+1}(1 + K(\gamma))^{n+1}}{(n+1)!} t^{n+1} < +\infty,$$

it follows from the Weierstrass criterion that $X_t^{(n)}(\gamma)$ converges uniformly on $[0, T]$, whose limit is denoted by $X_t(\gamma)$. Then X_t is just the solution of the uncertain differential equation (9.20). Next, we prove the uniqueness of the solution under the given conditions. Assume that X_t and X_t^* are two solutions of the multi-dimensional uncertain differential equation (9.20) with the same initial value X_0. By the Grönwall's inequality, we obtain

$$|X_t(\gamma) - X_t^*(\gamma)| \le 0, \quad \forall \gamma \in \Gamma$$

which means $X_t = X_t^*$ almost surely. The uniqueness of the solution is verified. The theorem is proved.

Stability Theorem

Definition 9.6 (*Su et al. [62]*) A multi-dimensional uncertain differential equation

$$dX_t = f(t, X_t)dt + g(t, X_t)dC_t \tag{9.24}$$

is said to be stable in measure if for any two solutions X_t and Y_t with different initial values X_0 and Y_0, we have

$$\lim_{|X_0 - Y_0| \to 0} \mathcal{M} \left\{ \sup_{t \ge 0} |X_t - Y_t| \le \varepsilon \right\} = 1 \tag{9.25}$$

for any given number $\varepsilon > 0$.

Example 9.13 Consider an m-dimensional uncertain differential equation

$$dX_t = U_t dt + V_t dC_t, \tag{9.26}$$

where U_t is an m-dimensional time integrable function, V_t is an $m \times n$ Liu integrable matrix function, and C_t is an n-dimensional canonical Liu process. Since its solutions with different initial values X_0 and Y_0 are

$$X_t = X_0 + \int_0^t U_s ds + \int_0^t V_s dC_s,$$

$$Y_t = Y_0 + \int_0^t U_s ds + \int_0^t V_s dC_s,$$

respectively, we have

$$\sup_{t \ge 0} |X_t - Y_t| = |X_0 - Y_0|$$

almost surely. Then

$$\lim_{|X_0-Y_0|\to 0} \mathcal{M}\left\{\sup_{t\geq 0}|X_t - Y_t| \leq \varepsilon\right\} = \lim_{|X_0-Y_0|\to 0} \mathcal{M}\{|X_0 - Y_0| \leq \varepsilon\} = 1,$$

and the m-dimensional uncertain differential equation (9.26) is stable in measure.

Example 9.14 Consider a 2-dimensional uncertain differential equation

$$dX_t = \begin{pmatrix} 1 & 0 \\ 0 & 1 \end{pmatrix} X_t dt + \begin{pmatrix} 1 & 0 \\ 0 & 1 \end{pmatrix} X_t dC_t \tag{9.27}$$

where C_t is a canonical Liu process. Since its two solutions with different initial values X_0 and Y_0 are

$$X_t = \begin{pmatrix} \exp(t + C_t) & 0 \\ 0 & \exp(t + C_t) \end{pmatrix} \cdot X_0,$$

$$Y_t = \begin{pmatrix} \exp(t + C_t) & 0 \\ 0 & \exp(t + C_t) \end{pmatrix} \cdot Y_0,$$

respectively, we have

$$\sup_{t\geq 0}|X_t(\gamma) - Y_t(\gamma)|$$

$$= \sup_{t\geq 0}\left|\begin{pmatrix} \exp(t + C_t(\gamma)) & 0 \\ 0 & \exp(t + C_t(\gamma)) \end{pmatrix} \cdot (X_0 - Y_0)\right|$$

$$\leq \sup_{t\geq 0}\left|\begin{pmatrix} \exp(t + C_t(\gamma)) & 0 \\ 0 & \exp(t + C_t(\gamma)) \end{pmatrix}\right| \cdot |X_0 - Y_0|$$

$$= \sup_{t\geq 0}\exp(t + C_t(\gamma)) \cdot |X_0 - Y_0| = +\infty$$

when $C_t(\gamma) \geq 0$ for all $t \in \Re$. Then

$$\lim_{|X_0-Y_0|\to 0} \mathcal{M}\left\{\sup_{t\geq 0}|X_t - Y_t| \leq \varepsilon\right\} \leq 1 - \mathcal{M}\{C_t \geq 0, \forall t\} = 1/2,$$

and the 2-dimensional uncertain differential equation (9.27) is not stable in measure.

Theorem 9.14 (Su et al. [62]) *Assume the multi-dimensional uncertain differential equation*

$$dX_t = f(t, X_t)dt + g(t, X_t)dC_t \tag{9.28}$$

has a unique solution for each given initial value. Then it is stable in measure if the coefficients $f(t, x)$ and $g(t, x)$ satisfy the strong Lipschitz condition

$$|f(t, x) - f(t, y)| + |g(t, x) - g(t, y)| \leq L_t|x - y|, \quad \forall x, y \in \Re^m, t \geq 0 \tag{9.29}$$

where L_t is some positive function satisfying

$$\int_0^{+\infty} L_t dt < +\infty. \tag{9.30}$$

Proof The proof of this theorem is similar to that of Theorem 6.8, so here it is introduced briefly. Let X_t and Y_t be the solutions of the multi-dimensional uncertain differential equation (9.28) with different initial values X_0 and Y_0, respectively. Then by the strong Lipschitz condition and the Grönwall's inequality, we have

$$\sup_{t \geq 0} |X_t - Y_t| \leq |X_0 - Y_0| \cdot \exp\left((1 + K)\int_0^{+\infty} L_s ds\right)$$

almost surely, where K is a nonnegative uncertain variable such that

$$\lim_{x \to \infty} \mathcal{M}\{\gamma \in \Gamma \mid K(\gamma) \leq x\} = 1$$

by Theorem 9.2. Note that for any given $\epsilon > 0$, there exists a real number H such that $\mathcal{M}\{\gamma \mid K(\gamma) \leq H\} \geq 1 - \epsilon$. Thus

$$\mathcal{M}\left\{\sup_{t \geq 0} |X_t - Y_t| \leq \varepsilon\right\} > 1 - \epsilon$$

as long as

$$|X_0 - Y_0| \leq \exp\left(-(1 + H)\int_0^{+\infty} L_s ds\right)\varepsilon.$$

In other words,

$$\lim_{|X_0 - Y_0| \to 0} \mathcal{M}\left\{\sup_{t \geq 0} |X_t - Y_t| \leq \varepsilon\right\} = 1,$$

and the uncertain differential equation (9.28) is stable in measure. The theorem is proved.

Chapter 10
High-Order Uncertain Differential Equation

High-order uncertain differential equation is a type of uncertain differential equations involving the high-order derivatives of uncertain processes. This chapter introduces high-order uncertain differential equation including its equivalent transformation to a multi-dimensional uncertain differential equation, Yao formula, numerical method, and existence and uniqueness theorem.

10.1 High-Order Uncertain Differential Equation

Definition 10.1 Suppose that C_t is a canonical Liu process, and $f(t, x_1, x_2, \ldots, x_n)$ and $g(t, x_1, x_2, \ldots, x_n)$ are two measurable functions. Then

$$\frac{d^n X_t}{dt^n} = f\left(t, X_t, \frac{dX_t}{dt}, \ldots, \frac{d^{n-1}X_t}{dt^{n-1}}\right) + g\left(t, X_t, \frac{dX_t}{dt}, \ldots, \frac{d^{n-1}X_t}{dt^{n-1}}\right)\frac{dC_t}{dt} \quad (10.1)$$

is called an n-order uncertain differential equation. An uncertain process that satisfies (10.1) identically at each time t is called a solution of the high-order uncertain differential equation.

Remark 10.1 When $n = 1$, the n-order uncertain differential equation (10.1) degenerates to the uncertain differential equation

$$dX_t = f(t, X_t)dt + g(t, X_t)dC_t.$$

Example 10.1 The 2-order uncertain differential equation

$$\frac{d^2 X_t}{dt^2} = \exp\left(-\frac{dX_t}{dt}\right) + \exp\left(-\frac{dX_t}{dt}\right) \cdot \frac{dC_t}{dt}, \quad X_0 = 0, \quad \left.\frac{dX_t}{dt}\right|_{t=0} = 0$$

© Springer-Verlag Berlin Heidelberg 2016
K. Yao, *Uncertain Differential Equations*,
Springer Uncertainty Research, DOI 10.1007/978-3-662-52729-0_10

has a solution

$$X_t = \int_0^t \ln(1 + s + C_s)\,ds.$$

Definition 10.2 The n-order uncertain differential equation in the form of

$$\frac{d^n X_t}{dt^n} = \left(u_{0t} + u_{1t}X_t + \cdots + u_{nt}\frac{d^{n-1}X_t}{dt^{n-1}} \right)$$
$$+ \left(v_{0t} + v_{1t}X_t + \cdots + v_{nt}\frac{d^{n-1}X_t}{dt^{n-1}} \right)\frac{dC_t}{dt} \qquad (10.2)$$

is called a linear n-order uncertain differential equation.

Remark 10.2 In a linear high-order uncertain differential equation, the coefficients $f(t, x_1, x_2, \ldots, x_n)$ and $g(t, x_1, x_2, \ldots, x_n)$ are linear functions of $x_1, x_2, \ldots, x_n$, i.e.,

$$f(t, x_1, x_2, \ldots, x_n) = u_{0t} + u_{1t}x_1 + \cdots + u_{nt}x_n,$$

$$g(t, x_1, x_2, \ldots, x_n) = v_{0t} + v_{1t}x_1 + \cdots + v_{nt}x_n.$$

Example 10.2 The linear 2-order uncertain differential equation

$$\frac{d^2 X_t}{dt^2} = \frac{dX_t}{dt} + \frac{dX_t}{dt} \cdot \frac{dC_t}{dt}, \quad X_0 = 0, \left. \frac{dX_t}{dt} \right|_{t=0} = 1$$

has a solution

$$X_t = \int_0^t \exp(s + C_s)\,ds.$$

10.2   Equivalent Transformation

For the n-order uncertain differential equation (10.1), write

$$X_{1t} = X_t, X_{2t} = \frac{dX_t}{dt}, \ldots, X_{nt} = \frac{d^{n-1}X_t}{dt^{n-1}}.$$

Then we have

$$
\begin{cases}
dX_{1t} = X_{2t}dt \\
dX_{2t} = X_{3t}dt \\
\quad\vdots \\
dX_{n-1,t} = X_{nt}dt \\
dX_{nt} = f(t, X_{1t}, X_{2t}, \ldots, X_{nt})dt + g(t, X_{1t}, X_{2t}, \ldots, X_{nt})dC_t.
\end{cases}
$$

In other words, the n-order uncertain differential equation (10.1) can be transformed into an n-dimensional uncertain differential equation

$$
dX_t = f(t, X_t)dt + g(t, X_t)dC_t \tag{10.3}
$$

where

$$
X_t = \begin{pmatrix} X_{1t} \\ \vdots \\ X_{n-1,t} \\ X_{nt} \end{pmatrix} = \begin{pmatrix} X_t \\ \vdots \\ \dfrac{d^{n-2}X_t}{dt^{n-2}} \\ \dfrac{d^{n-1}X_t}{dt^{n-1}} \end{pmatrix},
$$

$$
f(t, X_t) = \begin{pmatrix} X_{2t} \\ \vdots \\ X_{n,t} \\ f(t, X_{1t}, \ldots, X_{nt}) \end{pmatrix}, \quad g(t, X_t) = \begin{pmatrix} 0 \\ \vdots \\ 0 \\ g(t, X_{1t}, \ldots, X_{nt}) \end{pmatrix}.
$$

Particularly, the linear n-order uncertain differential equation (10.2) can be transformed into a linear n-dimensional uncertain differential equation

$$
dX_t = (U_{1t}X_t + U_{2t})dt + (V_{1t}X_t + V_{2t})dC_t
$$

where

$$
U_{1t} = \begin{pmatrix} 0 & 1 & \cdots & 0 \\ \vdots & \vdots & \ddots & \vdots \\ 0 & 0 & \cdots & 1 \\ u_{1t} & u_{2t} & \cdots & u_{nt} \end{pmatrix}, \quad U_{2t} = \begin{pmatrix} 0 \\ \vdots \\ 0 \\ u_{0t} \end{pmatrix},
$$

$$
V_{1t} = \begin{pmatrix} 0 & 0 & \cdots & 0 \\ \vdots & \vdots & \ddots & \vdots \\ 0 & 0 & \cdots & 0 \\ v_{1t} & v_{2t} & \cdots & v_{nt} \end{pmatrix}, \quad V_{2t} = \begin{pmatrix} 0 \\ \vdots \\ 0 \\ v_{0t} \end{pmatrix}.
$$

10.3   Yao Formula

Theorem 10.1   (Yao Formula) *The solution X_t of a high-order uncertain differential equation*

$$\frac{d^n X_t}{dt^n} = f\left(t, X_t, \frac{dX_t}{dt}, \ldots, \frac{d^{n-1}X_t}{dt^{n-1}}\right) + g\left(t, X_t, \frac{dX_t}{dt}, \ldots, \frac{d^{n-1}X_t}{dt^{n-1}}\right) \frac{dC_t}{dt} \quad (10.4)$$

is a contour process with an α-path X_t^α that solves the corresponding high-order ordinary differential equation

$$\frac{d^n X_t^\alpha}{dt^n} = f\left(t, X_t^\alpha, \frac{dX_t^\alpha}{dt}, \ldots, \frac{d^{n-1}X_t^\alpha}{dt^{n-1}}\right) + \left| g\left(t, X_t^\alpha, \frac{dX_t^\alpha}{dt}, \ldots, \frac{d^{n-1}X_t^\alpha}{dt^{n-1}}\right)\right| \Phi^{-1}(\alpha)$$
$$(10.5)$$

where

$$\Phi^{-1}(\alpha) = \frac{\sqrt{3}}{\pi} \ln \frac{\alpha}{1-\alpha} \quad (10.6)$$

is the inverse uncertainty distribution of standard normal uncertain variables. In other words,

$$\mathcal{M}\{X_t \le X_t^\alpha, \forall t\} = \alpha, \quad (10.7)$$

$$\mathcal{M}\{X_t > X_t^\alpha, \forall t\} = 1 - \alpha. \quad (10.8)$$

Proof Given $\alpha \in (0, 1)$, we divide the time interval into two parts

$$T^+ = \left\{ t \mid g\left(t, X_t^\alpha, \frac{dX_t^\alpha}{dt}, \ldots, \frac{d^{n-1}X_t^\alpha}{dt^{n-1}}\right) \ge 0\right\},$$

$$T^- = \left\{ t \mid g\left(t, X_t^\alpha, \frac{dX_t^\alpha}{dt}, \ldots, \frac{d^{n-1}X_t^\alpha}{dt^{n-1}}\right) < 0\right\}.$$

On the one hand, define

$$\Lambda_1^+ = \left\{ \gamma \left| \frac{dC_t(\gamma)}{dt}\right. \le \Phi^{-1}(\alpha) \text{ for any } t \in T^+\right\},$$

$$\Lambda_1^- = \left\{ \gamma \left| \frac{dC_t(\gamma)}{dt}\right. \ge \Phi^{-1}(1 - \alpha) \text{ for any } t \in T^-\right\}.$$

Noting that T^+ and T^- are disjoint sets and C_t is an independent increment uncertain process, we get

$$\mathcal{M}\{\Lambda_1^+\} = \alpha, \quad \mathcal{M}\{\Lambda_1^-\} = \alpha, \quad \mathcal{M}\{\Lambda_1^+ \cap \Lambda_1^-\} = \alpha.$$

For any $\gamma \in \Lambda_1^+ \cap \Lambda_1^-$, since

$$g\left(t, X_t(\gamma), \frac{dX_t(\gamma)}{dt}, \ldots, \frac{d^{n-1}X_t(\gamma)}{dt^{n-1}}\right) \frac{dC_t}{dt} \leq \left| g\left(t, X_t^\alpha, \frac{dX_t^\alpha}{dt}, \ldots, \frac{d^{n-1}X_t^\alpha}{dt^{n-1}}\right) \right| \Phi^{-1}(\alpha), \quad \forall t,$$

we have

$$X_t(\gamma) \leq X_t^\alpha, \quad \forall t$$

according to the comparison theorems of ordinary differential equations. Then

$$\mathcal{M}\{X_t \leq X_t^\alpha, \ \forall t\} \geq \mathcal{M}\{\Lambda_1^+ \cap \Lambda_1^-\} = \alpha. \tag{10.9}$$

On the other hand, define

$$\Lambda_2^+ = \left\{ \gamma \left| \frac{dC_t(\gamma)}{dt} \right. > \Phi^{-1}(\alpha) \text{ for any } t \in T^+ \right\},$$

$$\Lambda_2^- = \left\{ \gamma \left| \frac{dC_t(\gamma)}{dt} \right. < \Phi^{-1}(1-\alpha) \text{ for any } t \in T^- \right\}.$$

Noting that T^+ and T^- are disjoint sets and C_t is an independent increment uncertain process, we get

$$\mathcal{M}\{\Lambda_2^+\} = 1 - \alpha, \quad \mathcal{M}\{\Lambda_2^-\} = 1 - \alpha, \quad \mathcal{M}\{\Lambda_2^+ \cap \Lambda_2^-\} = 1 - \alpha.$$

For any $\gamma \in \Lambda_2^+ \cap \Lambda_2^-$, since

$$g\left(t, X_t(\gamma), \frac{dX_t(\gamma)}{dt}, \ldots, \frac{d^{n-1}X_t(\gamma)}{dt^{n-1}}\right) \frac{dC_t}{dt} > \left| g\left(t, X_t^\alpha, \frac{dX_t^\alpha}{dt}, \ldots, \frac{d^{n-1}X_t^\alpha}{dt^{n-1}}\right) \right| \Phi^{-1}(\alpha), \quad \forall t,$$

we have

$$X_t(\gamma) > X_t^\alpha, \quad \forall t$$

according to the comparison theorems of ordinary differential equations. Then

$$\mathcal{M}\{X_t > X_t^\alpha, \forall t\} \geq \mathcal{M}\{\Lambda_2^+ \cap \Lambda_2^-\} = 1 - \alpha. \tag{10.10}$$

Since

$$\mathcal{M}\{X_t \leq X_t^\alpha, \forall t\} + \mathcal{M}\{X_t > X_t^\alpha, \forall t\} \leq 1,$$

we have

$$\mathcal{M}\{X_t \leq X_t^\alpha, \forall t\} = \alpha, \quad \mathcal{M}\{X_t > X_t^\alpha, \forall t\} = 1 - \alpha$$

from Inequalities (10.9) and (10.10). The theorem is proved.

Remark 10.3 According to Chap. 4, the inverse uncertainty distribution, expected value, extreme value, and time integral of the solution of a high-order uncertain differential equation could all be obtained via the α-paths.

Example 10.3 The solution X_t of the linear 2-order uncertain differential equation

$$\frac{d^2X_t}{dt^2} = \frac{dX_t}{dt} + \frac{dC_t}{dt}, \quad X_0 = 0, \quad \frac{dX_t}{dt}\bigg|_{t=0} = 0$$

is a contour process with an α-path

$$X_t^\alpha = (\exp(t) - t - 1) \cdot \frac{\sqrt{3}}{\pi} \ln \frac{\alpha}{1 - \alpha}.$$

Example 10.4 The solution X_t of the linear 2-order uncertain differential equation

$$\frac{d^2X_t}{dt^2} = \frac{dX_t}{dt} + \frac{dX_t}{dt} \cdot \frac{dC_t}{dt}, \quad X_0 = 0, \quad \frac{dX_t}{dt}\bigg|_{t=0} = 1$$

is a contour process with an α-path

$$X_t^\alpha = \frac{\exp\left((1 + \Phi^{-1}(\alpha))t - 1\right)}{1 + \Phi^{-1}(\alpha)}$$

where

$$\Phi^{-1}(\alpha) = \frac{\sqrt{3}}{\pi} \ln \frac{\alpha}{1 - \alpha}.$$

Example 10.5 The solution X_t of the 2-order nonlinear uncertain differential equation

$$\frac{d^2X_t}{dt^2} = \exp\left(-\frac{dX_t}{dt}\right) + \exp\left(-\frac{dX_t}{dt}\right) \cdot \frac{dC_t}{dt}, \quad X_0 = 0, \quad \frac{dX_t}{dt}\bigg|_{t=0} = 0$$

is a contour process with an α-path

$$X_t^\alpha = \frac{1 + \left(1 + \Phi^{-1}(\alpha)\right)t}{1 + \Phi^{-1}(\alpha)} \cdot \ln\left(1 + \left(1 + \Phi^{-1}(\alpha)\right)t\right) - t$$

where

$$\Phi^{-1}(\alpha) = \frac{\sqrt{3}}{\pi} \ln \frac{\alpha}{1 - \alpha}.$$

10.4 Numerical Method

For a general high-order uncertain differential equation, it is difficult or impossible to find its analytic solution. Even if the analytic solution is available, sometimes we cannot get its uncertainty distribution, extreme value, or time integral. Alternatively, Yao Formula (Theorem 10.1) provides a numerical method to solve a high-order uncertain differential equation via the α-paths, whose procedure is designed as follows.

Step 1   Fix α on $(0, 1)$.
Step 2   Solve the high-order ordinary differential equation

$$\frac{d^n X_t^\alpha}{dt^n} = f\left(t, X_t^\alpha, \frac{dX_t^\alpha}{dt}, \ldots, \frac{d^{n-1}X_t^\alpha}{dt^{n-1}}\right) + \left| g\left(t, X_t^\alpha, \frac{dX_t^\alpha}{dt}, \ldots, \frac{d^{n-1}X_t^\alpha}{dt^{n-1}}\right) \right| \Phi^{-1}(\alpha) \tag{10.11}$$

by a numerical method where

$$\Phi^{-1}(\alpha) = \frac{\sqrt{3}}{\pi} \ln \frac{\alpha}{1 - \alpha}.$$

Step 3   Obtain the α-path.

Remark 10.4 The high-order ordinary uncertain differential equation (10.11) could be transformed equivalently into a system of ordinary differential equations by using the method of changing variables. Write

$$X_{1t}^\alpha = X_t^\alpha, X_{2t}^\alpha = \frac{dX_t^\alpha}{dt}, \ldots, X_{nt}^\alpha = \frac{d^{n-1}X_t^\alpha}{dt^{n-1}}.$$

Then we have

$$
\begin{cases}
dX_{1t}^{\alpha} = X_{2t}^{\alpha} dt \\
dX_{2t}^{\alpha} = X_{3t}^{\alpha} dt \\
\qquad \vdots \\
dX_{n-1,t}^{\alpha} = X_{nt}^{\alpha} dt \\
dX_{nt}^{\alpha} = f\left(t, X_{1t}^{\alpha}, X_{2t}^{\alpha}, \ldots, X_{nt}^{\alpha}\right) dt + \left| g\left(t, X_{1t}^{\alpha}, X_{2t}^{\alpha}, \ldots, X_{nt}^{\alpha}\right) \right| \Phi^{-1}(\alpha) dt.
\end{cases}
$$

This system of ordinary differential equations could be solved numerically using the Euler scheme

$$
\begin{cases}
X_{1,(i+1)h}^{\alpha} = X_{1,ih}^{\alpha} + X_{2,ih}^{\alpha} h \\
X_{2,(i+1)h}^{\alpha} = X_{2,ih}^{\alpha} + X_{3,ih}^{\alpha} h \\
\qquad \vdots \\
X_{n-1,(i+1)h}^{\alpha} = X_{n-1,ih}^{\alpha} + X_{n,ih}^{\alpha} h \\
X_{n,(i+1)h}^{\alpha} = X_{n,ih}^{\alpha} + f\left(t, X_{1,ih}^{\alpha}, X_{2,ih}^{\alpha}, \ldots, X_{n,ih}^{\alpha}\right) h + \left| g\left(t, X_{1,ih}^{\alpha}, X_{2,ih}^{\alpha}, \ldots, X_{n,ih}^{\alpha}\right) \right| \Phi^{-1}(\alpha) h.
\end{cases}
$$

Consider the solution of a high-order uncertain differential equation on the interval $[0, T]$. Fix N_1 and N_2 as some large enough integers. Let α change from ϵ to $1 - \epsilon$ with the step $\epsilon = 1/N_1$. By using the above Steps 1–3 with the step $h = T/N_2$, we obtain a spectrum of α-paths in discrete form $X_{ih}^{j\epsilon}$, $i = 1, 2, \ldots, N_2, j = 1, 2, \ldots, N_1$. Then the uncertain variable X_T has an inverse uncertainty distribution $X_{N_2 h}^{j\epsilon}$ in the discrete form and has an expected value

$$
E[X_T] = \sum_{j=1}^{N_1} X_{N_2 h}^{j\epsilon} / N_1.
$$

The supremum value

$$
\sup_{0 \le t \le T} X_t
$$

has an inverse uncertainty distribution

$$
\sup_{0 \le i \le N_2} X_{ih}^{j\epsilon}
$$

in the discrete form and has an expected value

$$
E\left[\sup_{0 \le t \le T} X_t \right] = \sum_{j=1}^{N_1} \left(\sup_{0 \le i \le N_2} X_{ih}^{j\epsilon} \right) / N_1.
$$

The infimum value

$$\inf_{0 \le t \le T} X_t$$

has an inverse uncertainty distribution

$$\inf_{0 \le i \le N_2} X_{ih}^{j\epsilon}$$

in the discrete form and has an expected value

$$E\left[\inf_{0 \le t \le T} X_t\right] = \sum_{j=1}^{N_1} \left(\inf_{0 \le i \le N_2} X_{ih}^{j\epsilon}\right) / N_1.$$

The time integral

$$\int_0^T X_t dt$$

has an inverse uncertainty distribution

$$\sum_{i=1}^{N_2} X_{ih}^{j\epsilon} / N_2$$

in the discrete form and has an expected value

$$E\left[\int_0^T X_t dt\right] = \sum_{j=1}^{N_1} \sum_{i=1}^{N_2} X_{ih}^{j\epsilon} / (N_2 N_1).$$

Example 10.6 Consider a 2-order uncertain differential equation

$$\frac{d^2 X_t}{dt^2} = \sin\left(\frac{dX_t}{dt}\right) + \cos(X_t)\frac{dC_t}{dt}, \quad X_0 = 1, \quad \frac{dX_t}{dt}\bigg|_{t=0} = 0.$$

We have

$$E[X_1] = 0.9749, \quad E\left[\sup_{0 \le t \le 1} X_t\right] = 1.1319,$$

$$E\left[\inf_{0 \le t \le 1} X_t\right] = 0.8430, \quad E\left[\int_0^1 X_t dt\right] = 0.9957.$$

Example 10.7 Consider a 2-order uncertain differential equation

$$\frac{d^2 X_t}{dt^2} = \sin(t + X_t) + \cos\left(t - \frac{dX_t}{dt}\right)\frac{dC_t}{dt}, \quad X_0 = 0, \quad \frac{dX_t}{dt}\bigg|_{t=0} = 0.$$

We have

$$E[X_1] = 0.1982, \quad E\left[\sup_{0 \le t \le 1} X_t\right] = 0.2854,$$

$$E\left[\inf_{0 \le t \le 1} X_t\right] = -0.0887, \quad E\left[\int_0^1 X_t dt\right] = 0.0426.$$

Example 10.8 Consider a 2-order uncertain differential equation

$$\frac{d^2 X_t}{dt^2} = \left(t - \frac{dX_t}{dt}\right) + (t + X_t^2)\frac{dC_t}{dt}, \quad X_0 = 1, \quad \frac{dX_t}{dt}\bigg|_{t=0} = 1.$$

We have

$$E[X_1] = 1.9258, \quad E\left[\sup_{0 \le t \le 1} X_t\right] = 1.9842,$$

$$E\left[\inf_{0 \le t \le 1} X_t\right] = 0.9744, \quad E\left[\int_0^1 X_t dt\right] = 1.4309.$$

10.5 Existence and Uniqueness Theorem

Theorem 10.2 *The n-order uncertain differential equation*

$$\frac{d^n X_t}{dt^n} = f\left(t, X_t, \frac{dX_t}{dt}, \ldots, \frac{d^{n-1} X_t}{dt^{n-1}}\right) + g\left(t, X_t, \frac{dX_t}{dt}, \ldots, \frac{d^{n-1} X_t}{dt^{n-1}}\right)\frac{dC_t}{dt} \qquad (10.12)$$

has a unique solution if for any $(x_1, \ldots, x_n)$, $(y_1, \ldots, y_n) \in \Re^n$ *and* $t \ge 0$, *the coefficients* $f(t, x_1, x_2, \ldots, x_n)$ *and* $g(t, x_1, x_2, \ldots, x_n)$ *satisfy the linear growth condition*

$$|f(t, x_1, \ldots, x_n)| + |g(t, x_1, \ldots, x_n)| \le L\left(1 + \bigvee_{i=1}^{n} |x_i|\right) \qquad (10.13)$$

and the Lipschitz condition

$$|f(t, x_1, \ldots, x_n) - f(t, y_1, \ldots, y_n)|$$

$$+ |g(t, x_1, \ldots, x_n) - g(t, y_1, \ldots, y_n)| \leq L \cdot \bigvee_{i=1}^{n} |x_i - y_i| \qquad (10.14)$$

for some constant L.

Proof Note that the high-order uncertain differential equation (10.12) can be transformed equivalently into the multi-dimensional uncertain differential equation (10.3). On the one hand, we have

$$|f(t, x)| = \bigvee_{i=2}^{n} |x_i| \vee |f(t, x_1, \ldots, x_n)| \leq (L+1) \left(1 + \bigvee_{i=1}^{n} |x_i| \right)$$

and

$$|g(t, x)| = |g(t, x_1, \ldots, x_n)| \leq L \left(1 + \bigvee_{i=1}^{n} |x_i| \right).$$

Then

$$|f(t, x)| + |g(t, x)| \leq (2L+1) \left(1 + \bigvee_{i=1}^{n} |x_i| \right).$$

On the other hand, we have

$$|f(t, x) - f(t, y)| = \bigvee_{i=2}^{n} |x_i - y_i| \vee |f(t, x_1, \ldots, x_n) - f(t, y_1, \ldots, y_n)|$$

$$\leq (L+1) \cdot \bigvee_{i=1}^{n} |x_i - y_i|$$

and

$$|g(t, x) - g(t, y)| = |g(t, x_1, \ldots, x_n) - g(t, y_1, \ldots, y_n)| \leq L \cdot \bigvee_{i=1}^{n} |x_i - y_i|.$$

Then

$$|f(t, x) - f(t, y)| + |g(t, x) - g(t, y)| \leq (2L + 1) \cdot \bigvee_{i=1}^{n} |x_i - y_i|.$$

According to Theorem 9.13, the multi-dimensional uncertain differential equation (10.3) has a unique solution, so the high-order uncertain differential equation (10.12) has a unique solution, too. The theorem is proved.

References

1. Barbacioru IC (2010) Uncertainty functional differential equations for finance. Surv Math Appl 5:275–284
2. Black F, Scholes M (1973) The pricing of option and corporate liabilities. J Political Econ 81:637–654
3. Chen X, Liu B (2010) Existence and uniqueness theorem for uncertain differential equations. Fuzzy Opt Decis Making 9(1):69–81
4. Chen X (2011) American option pricing formula for uncertain financial market. Int J Oper Res 8(2):32–37
5. Chen X, Dai W (2011) Maximum entropy principle for uncertain variables. Int J Fuzzy Syst 13(3):232–236
6. Chen X, Kar S, Ralescu DA (2012) Cross-entropy measure of uncertain variables. Inf Sci 201:53–60
7. Chen X (2012) Variation analysis of uncertain stationary independent increment process. Eur J Oper Res 222(2):312–316
8. Chen X, Ralescu DA (2013) Liu process and uncertain calculus. J Uncertain Anal Appl 1: Article 3
9. Chen X, Liu YH, Ralescu DA (2013) Uncertain stock model with periodic dividends. Fuzzy Opt Decis Mak 12(1):111–123
10. Chen X, Gao J (2013) Uncertain term structure model of interest rate. Soft Comput 17(4):597–604
11. Chen X, Gao J (2013) Stability analysis of linear uncetain differential equations. Ind Eng Manage Syst 12(1):2–8
12. Chen X (2015) Uncertain calculus with finite variation processes. Soft Comput 19(10):2905–2912
13. Deng LB, Zhu YG (2012) Uncertain optimal control with jump. ICIC Express Lett Part B: Appl 3(2):419–424
14. Dai W, Chen X (2012) Entropy of function of uncertain variables. Math Comput Model 55(3–4):754–760
15. Dubois D, Prade H (1988) Possibility theory: an approach to computerized processing of uncertainty. Plenum, New York
16. Gao J (2013) Uncertain bimatrix game with applications. Fuzzy Opt Decis Mak 12(1):65–78
17. Gao J, Yao K (2015) Some concepts and theorems of uncertain random process. Int J Intell Syst 30(1):52–65
18. Gao X (2009) Some properties of continuous uncertain measure. Int J Uncertain Fuzz 17(3):419–426
19. Gao Y (2012) Existence and uniqueness theorem on uncertain differential equations with local Lipschitz condition. J Uncertain Syst 6(3):223–232

© Springer-Verlag Berlin Heidelberg 2016

K. Yao, *Uncertain Differential Equations*,

Springer Uncertainty Research, DOI 10.1007/978-3-662-52729-0

20. Gao Y, Yao K (2014) Continuous dependence theorems on solutions of uncertain differential equations. Appl Math Model 38:3031–3037
21. Gao Y, Gao R, Yang LX (2013) Analysis of order statistics of uncertain variables. J Uncertain Anal Appl 3: Article 1
22. Ge XT, Zhu YG (2012) Existence and uniqueness theorem for uncertain delay differential equations. J Comput Inf Syst 8(20):8341–8347
23. Ge XT, Zhu YG (2013) A necessary condition of optimality for uncertain optimal control problem. Fuzzy Opt Decis Mak 12(1):41–51
24. Ito K (1944) Stochastic integral. Proc Jpn Acad Ser A 20(8):519–524
25. Ito K (1951) On stochastic differential equations. Mem Am Math Soc 4:1–51
26. Iwamura K, Kageyama M (2012) Exact construction of Liu process. Appl Math Sci 6(58):2871–2880
27. Iwamura K, Xu YL (2013) Estimating the variance of the square of canonical process. Appl Math Sci 7(75):3731–3738
28. Jeffreys H (1961) Theory of probability. Oxford University Press, Oxford
29. Ji X, Ke H (2016) Almost sure stability for uncertain differential equation with jumps. Soft Comput 20(2):547–553
30. Ji X, Zhou J (2015) Option pricing for an uncertain stock model with jumps. Soft Comput 19(11):3323–3329
31. Ji X, Zhou J (2015) Multi-dimensional uncertain differential equation: existence and uniqueness of solution. Fuzzy Opt Decis Mak 14(4):477–491
32. Jiao DY, Yao K (2015) An interest rate model in uncertain environment. Soft Comput 19(3):775–780
33. Kahneman D, Tversky A (1979) Prospect theory: an analysis of decision under risk. Econometrica 47(2):263–292
34. Li SG, Peng J, Zhang B (2015) Multifactor uncertain differential equation. J Uncertain Anal Appl 3: Article 7
35. Li X, Liu B (2009) Hybrid logic and uncertain logic. J Uncertain Syst 3(2):83–94
36. Liu B (2007) Uncertainty theory, 2nd edn. Springer, Berlin
37. Liu B (2008) Fuzzy process, hybrid process and uncertain process. J Uncertain Syst 2(1):3–16
38. Liu B (2009) Some research problems in uncertainty theory. J Uncertain Syst 3(1):3–10
39. Liu B (2010) Uncertain risk analysis and uncertain reliability analysis. J Uncertain Syst 4(3):163–170
40. Liu B (2010) Uncertainty theory: a branch of mathematics for modeling human uncertainty. Springer, Berlin
41. Liu B (2012) Why is there a need for uncertainty theory? J Uncertain Syst 6(1):3–10
42. Liu B, Yao K (2012) Uncertain integral with respect to multiple canonical processes. J Uncertain Syst 6(4):250–255
43. Liu B (2013) Toward uncertain finance theory. J Uncertain Anal Appl 1: Article 1
44. Liu B (2013) Extreme value theorems of uncertain process with application to insurance risk model. Soft Comput 17(4):549–556
45. Liu B (2013) Polyrectangular theorem and independence of uncertain vectors. J Uncertain Anal Appl 1: Article 9
46. Liu B (2014) Uncertainty distribution and independence of uncertain processes. Fuzzy Opt Decis Mak 13(3):259–271
47. Liu B (2015) Uncertainty theory, 4th edn. Springer, Berlin
48. Liu HJ, Fei WY (2013) Neutral uncertain delay differential equations. Inf: Int Interdiscip J 16(2A):1225–1232
49. Liu HJ, Ke H, Fei WY (2014) Almost sure stability for uncertain differential equation. Fuzzy Opt Decis Mak 13(4):463–473
50. Liu Y (2013) Semi-linear uncertain differential equation with its analytic solution. Inf: Int Interdiscip J 16(2A):889–894
51. Liu YH, Ha MH (2010) Expected value of function of uncertain variables. J Uncertain Syst 4(3):181–186

52. Liu YH (2012) An analytic method for solving uncertain differential equations. J Uncertain Syst 6(4):244–249
53. Liu YH (2015) Uncertain currency model and currency option pricing. Int J Intell Syst 30(1):40–51
54. Øksendal B (2005) Stochastic differential equations, 6th edn. Springer, Berlin
55. Peng J, Yao K (2011) A new option pricing model for stocks in uncertainty markets. Int J Oper Res 8(2):18–26
56. Peng J (2013) Risk metrics of loss function for uncertain system. Fuzzy Opt Decis Mak 12(1):53–64
57. Peng ZX, Iwamura K (2010) A sufficient and necessary condition of uncertainty distribution. J Interdiscip Math 13(3):277–285
58. Peng ZX, Iwamura K (2012) Some properties of product uncertain measure. J Uncertain Syst 6(4):263–269
59. Sheng LX, Zhu YG, Hamalaonen T (2013) An uncertain optimal control with Hurwicz criterion. Appl Math Comput 224:412–421
60. Sheng YH, Wang CG (2014) Stability in p-th moment for uncertain differential equation. J Intell Fuzzy Syst 26(3):1263–1271
61. Sheng YH, Kar S (2015) Some results of moments of uncertain variable through inverse uncertainty distribution. Fuzzy Opt Decis Mak 14(1):57–76
62. Su TY, Wu HS, Zhou J (2015) Stability of multi-dimensional uncertain differential equation. Soft Comput. doi:10.1007/s00500-015-1788-0
63. Sun JJ, Chen X (2015) Asian option pricing formula for uncertain financial market. J Uncertain Anal Appl 3: Article 11
64. Tian JF (2011) Inequalities and mathematical properties of uncertain variables. Fuzzy Opt Decis Mak 10(4):357–368
65. Wang X, Ning YF, Moughal TA, Chen XM (2015) Adams-Simpson method for solving uncertain differential equation. Appl Math Comput 271:209–219
66. Wang XS, Peng ZX (2014) Method of moments for estimating uncertainty distributions. J Uncertain Anal Appl 2: Article 5
67. Xu XX, Zhu YG (2012) Uncertain bang-bang control for continuous time model. Cybern Syst 43(6):515–527
68. Yang XF, Ralescu DA (2015) Adams method for solving uncertain differential equations. Appl Math Comput 270:993–1003
69. Yang XF, Shen YY (2015) Runge–Kutta method for solving uncertain differential equations. J Uncertain Anal Appl 3: Article 17
70. Yang XF, Gao J, Kar S, Uncertain calculus with Yao process. http://orsc.edu.cn/online/150602.pdf
71. Yao K (2010) Expected value of lognormal uncertain variable. In: Proceedings of the first international conference on uncertainty theory, Urumchi, China, 11–19 August 2010, pp 241–243
72. Yao K (2012) Uncertain calculus with renewal process. Fuzzy Opt Decis Mak 11(3):285–297
73. Yao K, Li X (2012) Uncertain alternating renewal process and its application. IEEE Trans Fuzzy Syst 20(6):1154–1160
74. Yao K, Gao J, Gao Y (2013) Some stability theorems of uncertain differential equation. Fuzzy Opt Decis Mak 12(1):3–13
75. Yao K, Chen X (2013) A numerical method for solving uncertain differential equations. J Intell Fuzzy Syst 25(3):825–832
76. Yao K (2013) Extreme values and integral of solution of uncertain differential equation. J Uncertain Anal Appl 1: Article 2
77. Yao K (2013) A type of uncertain differential equations with analytic solution. J Uncertain Anal Appl 1: Article 8
78. Yao K, Ralescu DA (2013) Age replacement policy in uncertain environment. Iran J Fuzzy Syst 10(2):29–39

79. Yao K (2014) Multi-dimensional uncertain calculus with Liu process. J Uncertain Syst 8(4):244–254
80. Yao K (2015) A no-arbitrage theorem for uncertain stock model. Fuzzy Opt Decis Mak 14(2):227–242
81. Yao K, Qin ZF (2015) A modified insurance risk process with uncertainty. Insur: Math Econ 62:227–233
82. Yao K, Ke H, Sheng YH (2015) Stability in mean for uncertain differential equation. Fuzzy Opt Decis Mak 14(3):365–379
83. Yao K (2015) Uncertain differential equation with jumps. Soft Comput 19(7):2063–2069
84. Yao K (2015) A formula to calculate the variance of uncertain variable. Soft Comput 19(10):2947–2953
85. Yao K (2015) Uncertain contour process and its application in stock model with floating interest rate. Fuzzy Opt Decis Mak 14(4):399–424
86. You C (2009) Some convergence theorems of uncertain sequences. Math Comput Model 49(3–4):482–487
87. Wang Z (2013) Analytic solution for a general type of uncertain differential equation. Inf: Int Interdiscip J 16(2A):1003–1010
88. Yu XC (2012) A stock model with jumps for uncertain markets. Int J Uncertain Fuzz 20(3):421–432
89. Zhang CX, Guo CR (2014) Uncertain block replacement policy with no replacement at failure. J Intell Fuzzy Syst 27(4):1991–1997
90. Zhang TC, Chen X (2013) Multi-dimensional canonical process. Inf: Int Interdiscip J 16(2A): 1025–1030
91. Zhang XF, Ning YF, Meng GW (2013) Delayed renewal process with uncertain interarrival times. Fuzzy Opt Decis Mak 12(1):79–87
92. Zhang ZQ, Liu WQ (2014) Geometric average Asian option pricing for uncertain financial market. J Uncertain Syst 8(4):317–320
93. Zhu YG (2010) Uncertain optimal control with application to a portfolio selection model. Cybern Syst 41(7):535–547
94. Zhu YG (2012) Functions of uncertain variables and uncertain programming. J Uncertain Syst 6(4):278–288
95. Zhu YG (2015) Uncertain fractional differential equations and an interest rate model. Math Meth Appl Sci 38(15):3359–3368

Index

A
Almost sure stability, 77, 114
α-path, 29

C
Canonical Liu process, 39
Contour process, 29

E
Euler scheme, 62, 148
Event, 5
Expected value, 16

F
Fundamental theorem, 45, 101, 129

H
High-order ude, 141

I
Increment, 21
Independence, 9, 10, 22
Infimum process, 24
Integration by parts, 47, 103, 132
Inverse uncertainty distribution, 13

L
Liu integral, 41
Liu process, 44

M
Multi-dimensional integral, 125
Multi-dimensional process, 123, 129
Multi-dimensional ude, 133

O
Operational law, 11, 14

R
Renewal process, 95
Runge–Kutta scheme, 62

S
Sample continuity, 22
Sample path, 21
Stability in mean, 71
Stability in measure, 67, 111, 137
Stock model, 81, 86, 118
Supremum process, 23

T
Time integral, 25

U
Ude with jumps, 105
Uncertain differential equation, 49
Uncertain measure, 5
Uncertain process, 21
Uncertain variable, 8
Uncertain vector, 9vfill

© Springer-Verlag Berlin Heidelberg 2016
K. Yao, *Uncertain Differential Equations*,
Springer Uncertainty Research, DOI 10.1007/978-3-662-52729-0

Printed in the United States
By Bookmasters

Printed in the United States
By Bookmasters